REVISE EDEXCEL GCSE (9–1)
Combined Science
REVISION WORKBOOK
Foundation

Series Consultant: Harry Smith

Authors: Stephen Hoare, Nigel Saunders and Catherine Wilson

Notes from the publisher

While the publishers have made every attempt to ensure that advice on the qualification and its assessment is accurate, the official specification and associated assessment guidance materials are the only authoritative source of information and should always be referred to for definitive guidance.

Pearson examiners have not contributed to any sections in this resource relevant to examination papers for which they have responsibility.

Question difficulty

Look at this scale next to each exam-style question. It tells you how difficult the question is.

For the full range of Pearson revision titles across KS2, KS3, GCSE, Functional Skills, AS/A Level and BTEC visit:
www.pearsonschools.co.uk/revise

Contents

- - - - - - - - - - - - - - - - -

A small bit of small print: Edexcel
publishes Sample Assessment Material
and the Specification on its website.
This is the official content and this book
should be used in conjunction with it.
The questions have been written to help
you practise every topic in the book.
Remember: the real exam questions may
not look like this.

Published by Pearson Education Limited, 80 Strand, London, WC2R 0RL.

www.pearsonschoolsandfecolleges.co.uk

Copies of official specifications for all Pearson qualifications may be found on the website: qualifications.pearson.com

Text and illustrations © Pearson Education Limited 2017
Typeset and produced by Phoenix Photosetting
Illustrated by TechSet Ltd and Phoenix Photosetting
Cover illustration by Miriam Sturdee

The rights of Stephen Hoare, Nigel Saunders and Catherine Wilson to be identified as authors of this work has been asserted by them in accordance with the Copyright, Designs and Patents Act 1988.

First published 2017

20 19 18
10 9 8 7 6 5 4 3 2

British Library Cataloguing in Publication Data
A catalogue record for this book is available from the British Library

ISBN 978 1 292 13155 9

Printed in Slovakia by Neografia.

Acknowledgements
Content written by Allison Court, Ian Roberts, Damian Riddle, Julia Salter and Stephen Winrow-Campbell is included.

The authors and publisher would like to thank the following individuals and organisations for their kind permission to reproduce copyright material.

Photographs
(Key: b-bottom; c-centre; l-left; r-right; t-top)
Alamy Stock Photo: Bruce Boulton.co.uk 151; **NASA:** David R. Scott 163; **Pearson Education Ltd:** Oxford Designers & Illustrators Ltd 140; **Science Photo Library:** Biophoto Associates 003, Steve Gschmeissner 006, 013

All other images © Pearson Education

Notes from the publisher

1. While the publishers have made every attempt to ensure that advice on the qualification and its assessment is accurate, the official specification and associated assessment guidance materials are the only authoritative source of information and should always be referred to for definitive guidance.

Pearson examiners have not contributed to any sections in this resource relevant to examination papers for which they have responsibility.

2. Pearson has robust editorial processes, including answer and fact checks, to ensure the accuracy of the content in this publication, and every effort is made to ensure this publication is free of errors. We are, however, only human, and occasionally errors do occur. Pearson is not liable for any misunderstandings that arise as a result of errors in this publication, but it is our priority to ensure that the content is accurate. If you spot an error, please do contact us at resourcescorrections@pearson.com so we can make sure it is corrected.

Plant and animal cells

1 (a) Which of the following are found in both animal and plant cells?

☐ **A** cell membrane, nucleus, chloroplast

☐ **B** cell membrane, nucleus, ribosomes

☐ **C** cell wall, nucleus, ribosomes

☐ **D** cell wall, mitochondria, ribosomes   **(1 mark)**

> Look at the mark allocation for each question – here there is one mark so you need to put a cross in **one** box.

(b) Which of the following are found only in plant cells?

☐ **A** cell membrane, nucleus, chloroplast

☐ **B** cell membrane, vacuole, chloroplast

☐ **C** cell wall, chloroplast, vacuole

☐ **D** cytoplasm, chloroplast, vacuole   **(1 mark)**

> Always answer multiple-choice questions, even if you don't actually know the answer.

2 (a) Describe the function of mitochondria.

....

....

.... **(2 marks)**

(b) Explain why all plant cells contain mitochondria but only some contain chloroplasts.

> Chloroplasts need light to carry out photosynthesis. Use the function of a chloroplast to explain why you would not find them in certain cells, such as root cells.

....

....

....

.... **(2 marks)**

3 Describe the difference between the functions of a cell membrane and a cell wall.

> **Guided**

Cell membrane controls ....

....

....

.... **(2 marks)**

4 Enzymes are proteins made in cells. Pancreatic cells produce large amounts of enzymes but fat cells do not. Suggest an explanation for why pancreatic cells contain many more ribosomes than fat cells.

....

....

....

.... **(2 marks)**

Different kinds of cell

1 The genes in a bacterial cell are contained:

    ☐ **A** on a circular chromosome only

    ☐ **B** on plasmids only

    ☐ **C** on plasmids and a circular chromosome

    ☐ **D** ~~in the nucleus~~ | It cannot be D because bacteria do not have nuclei. | **(1 mark)**

2 The diagram shows a sperm cell and a bacterium. Note that the drawings are not to the same scale.

A ⎯

B

Sperm cell   Bacterium

(a) Name the structures labelled A and B in the diagram:

A ....

B .... **(2 marks)**

(b) Describe the function of each structure.

A ....

....

B ....

.... **(2 marks)**

3 Breathing can expose us to dust, dirt and bacteria.

Explain how cells in the lungs are specialised to protect us from these.

....

....

....

....

....

.... **(3 marks)**

Microscopes and magnification

1   Scientists use two types of microscope to examine cells: light microscopes and electron microscopes. Describe how these types of microscope are different.

Light microscopes magnify .... than electron microscopes.

The level of cell detail seen with an electron microscope is ....

because .... **(3 marks)**

2   The image shows an electron micrograph of part of a human liver cell.

(a) Explain why this is a eukaryotic cell.

....

....

....**(2 marks)**

mitochondrion —

nucleus —

2 µm

(b) Estimate the size of the following parts of the cell:

(i)   the nucleus

.... **(2 marks)**

(ii) the mitochondrion

.... **(2 marks)**

(c) Explain why it would be possible to see the nucleus clearly using a light microscope, but the mitochondria would be unclear.

....

....

.... **(3 marks)**

3   A scientist wants to study some bacteria that are 2.5 µm long. She can use either a light microscope (the one in the lab has a magnification of ×1000) or an electron microscope (the one in the lab next door has a magnification of ×100 000).

(a) Calculate the size of the magnified image of the bacteria seen with each type of microscope.

> Remember that $1 \, \mu m = 1 \times 10^{-6} \, m$ and do a reality check on your answer. The magnified image must be **bigger** than the bacteria. The image formed by the electron microscope must be **bigger** than that formed by the light microscope.

(3 marks)

(b) Explain which microscope would be better for her to use.

....

....

> State which is better **and** give a reason.

.... **(2 marks)**

3

Dealing with numbers

Guided

1   Give the following units in order of increasing size:

   metre   micrometre   millimetre   nanometre   picometre

   picometre .... metre   **(1 mark)**

Guided

2   Complete the table to convert the quantities to the units shown.

Quantity	Converted quantity	
0.005 nanometres	5	picometres
250 milligrams		grams
250 milligrams		kilograms
2.5 metres		millimetres

(4 marks)

Guided

3   For each of the following conversions, state whether it is true or false.

Conversion	True or false?
$0.000\,125\,\text{mm} = 0.125\,\mu\text{m}$	true
$150\,000\,\text{mg} = 0.015\,\text{kg}$	
$1\,\text{kg} = 10\,000\,000\,\mu\text{g}$	
$0.25\,\text{mm} = 2.5 \times 10^{2}\,\mu\text{m}$	

(4 marks)

4   Calculate for each of the following the actual size of the structure.

> 🖩 **Maths skills** $1\,\text{mm} = 1\,000\,\mu\text{m}$, and $1\,\text{mm} = 1\,000\,000\,\text{nm}$
> (check back on page 4 of the Revision Guide).

(a) a ribosome that measured 30.9 mm in an electron micrograph (magnification = ×1 000 000)

.... nm   **(2 marks)**

(b) a mitochondrion that measured 163 mm in an electron micrograph (magnification = ×250 000)

.... nm   **(2 marks)**

(c) a nucleus that measured 7.8 mm in a light microscope (magnification = ×800)

.... µm   **(2 marks)**

Practical skills Using a light microscope

1 (a) State the function of the following parts of a light microscope:

   (i)   the mirror

   .... **(1 mark)**

   (ii) the stage with clips

   .... **(1 mark)**

   (iii) the coarse focusing wheel

   .... **(1 mark)**

   (b) Give the reasons for the following precautions when using a light microscope.

   (i)   Never use the coarse focusing wheel with a high power objective.

   ....

   .... **(1 mark)**

   (ii) Never point the mirror directly at the Sun.

   ....

   .... **(1 mark)**

   (c) (i)   State an alternative light source that might be safer than the Sun.

   .... **(1 mark)**

> **Guided**

   (ii) State **two** other precautions that you should take when using a light microscope.

   precaution 1   Always start with the lowest power objective under the eyepiece.

   precaution 2   ....

   .... **(2 marks)**

2 You are observing a slide under high power but cannot see the part you need. Describe how you would bring the required part into view.

| Think about why you cannot see what you need and then the steps you must follow to find it. Remember some of the precautions you have to take. |

....

....

....

....

....

.... **(3 marks)**

Practical skills **Drawing labelled diagrams**

Guided

1 A student was given the slide below left and told to make a high power drawing to show cells in different stages of mitosis. His drawing is shown below right.

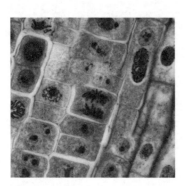

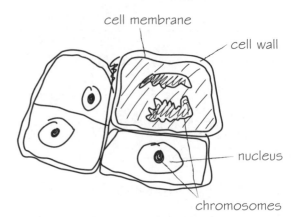

cell membrane

cell wall

nucleus

chromosomes

(a) Identify three faults with the student's drawing.

fault 1   The drawing is in pen rather than in ....

fault 2   ....

fault 3   ....   **(3 marks)**

(b) Draw your own labelled diagram of the slide above.

> Include outlines of all cells with more detail of cells showing different stages of mitosis. Try to show one of each stage.

(4 marks)

2 The student used a scale to measure the actual width of the field of view shown in the slide (above left) and found it was 0.113 mm. Calculate the magnification.

magnification = ...   **(3 marks)**

Enzymes

1 The enzyme invertase digests sucrose to glucose and fructose. Explain why invertase will not digest the sugar lactose.

> **Guided**

The shape of ....

matches the shape of ....

so .... cannot combine with ....

....

.... **(2 marks)**

2 The graph shows how the activity of an enzyme changes with temperature. The enzyme activity is greatest at the temperature labelled B.

(a) Give the optimum temperature for this enzyme.

.... **(1 mark)**

(b) Why does the enzyme activity decrease in the region labelled C?

☐ **A** Enzymes are killed at high temperatures.

☐ **B** The active site cannot change shape.

☐ **C** The substrate molecules move more slowly.

☐ **D** The active site breaks up and the enzyme is denatured.   **(1 mark)**

(c) Explain why the enzyme activity increases in the region labelled A.

> Think about the rate of collisions involving molecules as the temperature increases.

....

.... **(2 marks)**

3 Amylase is an enzyme that digests starch. Its optimum pH is about 7. Pepsin and trypsin are enzymes that digest proteins. Pepsin is produced in the stomach (pH 2). Trypsin is found in pancreatic juice (pH 8.6), which is released into the small intestine. The graph shows the effect of pH on the activity of these enzymes.

Use this information to explain why proteins are digested in the stomach and small intestine.

....

....

....

....

....

(3 marks)

7

Practical skills pH and enzyme activity

1 A student carried out an experiment to investigate the effect of pH on the activity of the enzyme trypsin using pieces of photographic film. Trypsin digests the protein in the film and causes the film to turn clear. Measuring the time it takes for the film to clear allows you to calculate the rate of reaction. The student used the apparatus shown.

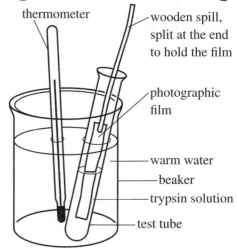

This procedure was repeated using trypsin solution at different pH values. The student's results are shown in the table.

pH	2	4	6	8	10
Time (min)	> 10	7.5	3.6	1.2	8.3
Rate/min	0	0.13			

Remember, rate = $\dfrac{1}{time}$

(a) Complete the table by calculating the rate of reaction at each pH. **(2 marks)**

(b) Complete the graph to show the effect of pH on the rate of reaction.

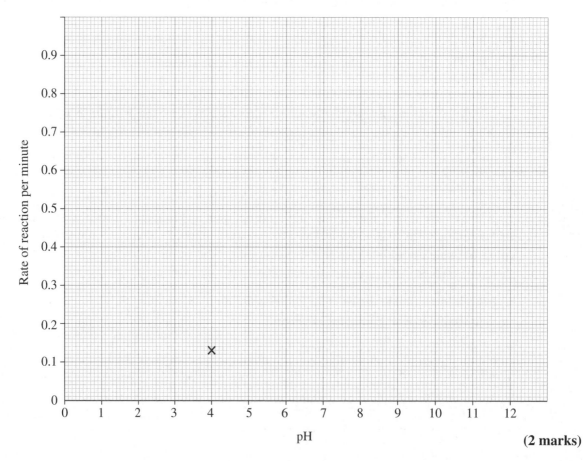

(2 marks)

(c) State **two** ways in which the experiment could be improved.

improvement 1 ....

improvement 2 .... **(2 marks)**

The importance of enzymes

Guided

1 Complete the following table.

Enzyme	Digests	Product(s)
amylase	starch	
lipase		
protease		amino acids

(3 marks)

2 (a) Explain why different digestive enzymes are needed in the digestive system.

....

....

.... **(2 marks)**

(b) Explain the importance of enzymes as biological catalysts in building the molecules needed in cells and tissues.

....

....

.... **(2 marks)**

3 Biological washing powders contain enzymes that help to break down food stains on clothes.

(a) Eggs are rich in protein. An egg fried in oil is spilled on a shirt, causing a stain.
Complete the table to show the enzyme or enzymes needed to remove this stain.
Place a tick (✓) in each correct box.

Enzyme	Needed to remove the stain? (✓)
amylase	
lipase	
protease	

(1 mark)

(b) Explain why biological washing powders work better below 40 °C.

> Think about what biological washing powders contain and what effect temperature might have.

....

....

....

.... **(3 marks)**

4 Give **one** similarity and **one** difference between digestion and synthesis in living organisms.

....

....

....

.... **(3 marks)**

9

Getting in and out of cells

1 Define diffusion.

....

....

....

.... **(2 marks)**

2 The table shows some features of two transport processes. Complete the table by placing a tick (✓) in each correct box to show the features of diffusion and of active transport.

Feature	Diffusion	Active transport
Involves the movement of particles	✓	✓
Requires energy		
Can happen across a partially permeable membrane		
Net movement down a concentration gradient		

(4 marks)

3 (a) Explain what is meant by the term **osmosis**.

> Make sure you use the terms 'water', 'partially permeable membrane' and 'movement' in your answer.

Osmosis is the net movement of .... across a

....

from a low ....

to a high .... **(4 marks)**

(b) The blood in the lung capillaries has a lower concentration of oxygen than the air. Oxygen moves from the air to the blood. Name the transport process involved, and give a reason that explains your answer.

....

....

....

....

....

.... **(2 marks)**

(c) Starch in food is digested to glucose. It is important that all the glucose produced is absorbed from the small intestine. Explain why this process requires energy.

....

....

....

.... **(2 marks)**

Practical skills   Osmosis in potatoes

> Guided

1 Describe how you would investigate osmosis in potatoes using potato pieces. You are provided with solutions of different sucrose concentrations. You should include at least **two** steps that you should use to ensure the accuracy of your results.

Cut pieces of potato, making sure ....

....

....

Remove from the solution, then ....

.... **(4 marks)**

> Guided

2 The table shows the results of an experiment to investigate osmosis in potatoes using different concentrations of sucrose.

Concentration (mol dm^{-3})	Initial mass (g)	Final mass (g)	Change in mass (g)	Percentage change in mass (%)
0	2.60	2.85		9.6
0.2	2.51	2.67	0.16	6.4
0.4	2.65	2.72	0.07	2.6
0.6	2.52	2.45	−0.07	−2.8
0.8	2.58	2.43	−0.15	

(a) Calculate the two missing values in the table. Use these values to complete the table.

2.85 – 2.60 = ... g

percentage change in mass = $\dfrac{-0.15}{2.58} \times 100 =$ .... %

(2 marks)

(b) Complete the graph to show the percentage change in mass against sucrose concentration.
(2 marks)

> **Maths skills** Draw a line of best fit. This can be curved or straight, depending on the data, but should ignore points that are clearly anomalies.

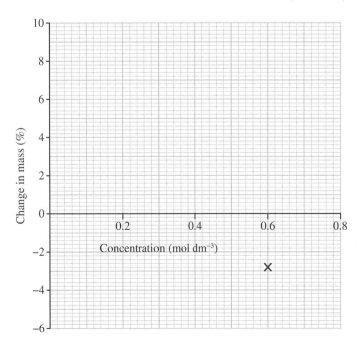

(c) Use the completed graph to estimate the concentration of the potato tissue.

....

.... **(1 mark)**

Extended response – Key concepts

The diagrams show a bacterial cell and a plant cell. The diagrams are not drawn to scale.

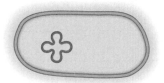

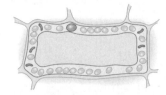

Bacterial cell   Plant cell

Compare the structures of these two cells, including subcellular structures and their functions.

> In your answer to this question, you need to think about:
> - the similarities between the cells
> - the differences between the cells.
>
> For each structure that you identify, remember to describe its function.
>
> It may help if you make a brief plan before you start writing.

....

....

....

....

....

....

....

....

....

....

....

....

....

....

....

.... **(6 marks)**

Mitosis

1 (a) Read the following statements about mitosis. Which statement is correct? Tick **one** box.

☐ **A** A parent cell divides to produce two genetically different diploid daughter cells.

☐ **B** A parent cell divides to produce two genetically identical diploid daughter cells.

☐ **C** A parent cell divides to produce four genetically identical haploid daughter cells.

☐ **D** A parent cell divides to produce four genetically different haploid cells. **(1 mark)**

(b) List the stages of mitosis in the order they happen, starting with interphase.

.... **(1 mark)**

2 Give **three** reasons why mitosis takes place.

> **Guided**

1 to produce new individuals by .... reproduction

2 ....

3 .... **(3 marks)**

3 The photograph shows a slide of cells from an onion root tip at different stages in mitosis.

(a) Name the two stages of mitosis labelled A and B.

A ....

B .... **(2 marks)**

(b) Give a reason for each answer in part (a).

A ....

B .... **(2 marks)**

Had a go ☐   Nearly there ☐   Nailed it! ☐

Cell growth and differentiation

1   (a)   Give the name of a fertilised egg in animals.

.... **(1 mark)**

   (b)   State the type of cell division that occurs after an egg is fertilised.

.... **(1 mark)**

2   Plant cells divide by mitosis.

   (a)   State the name of the type of plant tissue where mitosis occurs rapidly.

> Remember that plants grow when cells divide and when cells elongate.

.... **(1 mark)**

   (b)   Describe how plant cells increase in size following mitosis.

....

.... **(2 marks)**

3   (a)   Complete the table to show whether the different specialised cells are animal or plant cells.

Guided

Type of specialised cell	Animal or plant
sperm	animal
xylem	
ciliated cell	
root hair cell	
egg cell	

(3 marks)

   (b)   Give the name of one other type of specialised cell found in **plants** and one in **animals**.

plants ....

animals .... **(2 marks)**

4   Growth in animals happens over a particular period of the animal's lifespan. Growth happens through cell division and when cells in the animal differentiate.

Guided

   (a)   Explain what is meant by the term **differentiate**.

Cells become .... to perform ....

.... **(2 marks)**

   (b)   Give **one** reason that explains why cell differentiation is important in animals.

....

....

.... **(2 marks)**

Growth and percentile charts

1   A midwife will measure the growth of a baby in different ways. The graph shows some percentile charts for the head circumference measurement for young children.

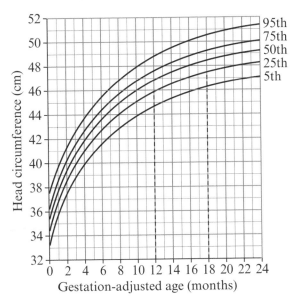

> Graphs like this sometimes look complicated – but remember that the curves are all labelled, so you can see what each one refers to.

> Note that dashed guidelines have been put in to help you answer part (b). These help to show how you get the reading for both measurements from the graph – you can then subtract one number from the other to get the final answer.

(a)   The median head circumference is described by the line where half the babies have a greater circumference, and half have the same or a smaller circumference. Which percentile curve shows the median rate of growth for babies?

☐ **A** 5th percentile   ☐ **B** 25th percentile   ☐ **C** 50th percentile   ☐ **D** 75th percentile

(1 mark)

(b)   Use the graph to calculate the change in head circumference for a baby that lies on the 25th percentile curve between 12 and 18 months old. Show your working.

change in circumference .... cm   **(2 marks)**

2   Growth in seedlings can be investigated by measuring the mass of seedlings of different ages.

> **Guided**

(a)   One seedling increased in mass from 12.75 g to 15.35 g over a period of 7 days. Calculate the percentage increase in mass for this seedling. Show your working.

15.35 – 12.75 = ...g

$\dfrac{..............}{12.75} \times 100 =$ .... %   **(2 marks)**

(b) Describe **one** other way you could measure the growth of the seedlings.

....

....

....

.... **(2 marks)**

15

Stem cells

1   Plants and animals have stem cells.

> **Guided**

   (a) What is a stem cell?

     ☐ **A** an undifferentiated cell

     ☐ **B** a specialised cell of an organism

> Option D cannot be correct because stem cells may be able to help conditions such as these.

     ☐ **C** a cell found only in embryo plants and animals

     ☐ **D** ~~a cell that causes Parkinson's disease and some types of blindness.~~ **(1 mark)**

   (b) (i)   Give the name of the tissue where plant stem cells are found.

      .... **(1 mark)**

     (ii) Name **two** places in a plant where you would find stem cells.

      .... and .... **(2 marks)**

2   (a) Describe **one** function of adult stem cells.

     ....

     .... **(1 mark)**

   (b) Describe **one** difference between an embryonic stem cell and a differentiated cell.

     ....

     .... **(1 mark)**

> To answer the following questions, think about what happens in a tissue transplant as well as what the different types of stem cell are capable of.

3   Parkinson's disease is caused by the death of some types of nerve cells in the brain.

   (a) Describe how embryonic stem cells could be used to treat Parkinson's disease.

     ....

     .... **(2 marks)**

   (b) IPSCs are stem cells produced by modifying a patient's own skin cells. IPSCs could also be used to treat Parkinson's disease in the future.

     (i)   Give **one** benefit of using IPSCs rather than embryonic stem cells for treating disease.

> Benefits could involve addressing ethical concerns or practical ones.

      .... **(1 mark)**

     (ii) Suggest **one** risk of using IPSCs.

      .... **(1 mark)**

Neurones

1   Draw **one** line from each type of neurone to its correct function.

    Neurone   **Function**

    | motor neurone | | carries impulses from one part of the central nervous system to another |

    | relay neurone | | carries impulses to the central nervous system |

    | sensory neurone | | carries impulses from the central nervous system to effectors |

                                                                                    (3 marks)

2   The diagram shows a sensory neurone.

    A ....

    B ....

    C ....

    D...

    E...

    F............................

    Label the parts A – F of the sensory neurone. Write your answers on the diagram.   **(3 marks)**

3   Explain how the structure of a motor neurone is related to its function.

    Guided

    The axon is long so it can...

    The axon has a myelin sheath which..

    The nerve ending transmits impulses to .... **(3 marks)**

4   The table shows the speed at which nerve impulses are carried along two types of neurone.

Type of neurone	Speed of transmission (m/s)
myelinated	25
unmyelinated	3

    (a) Explain why the speed of transmission is different in the two types of neurone.

    ....

    ....

    .... **(2 marks)**

    (b) In multiple sclerosis (MS), the myelin sheath surrounding neurones in the spinal cord is destroyed. Explain what effect this would have on the movement of a person with MS.

    ....

    ....

    .... **(2 marks)**

Responding to stimuli

1   Choose **three** words from the box to complete the sentence about the features of reflex actions.

innate	slow	automatic	conscious	learned	rapid

Reflex actions are ...., ..., and .... **(3 marks)**

2   The diagram shows a junction where neurone X meets neurone Y.

electrical impulse

axon of neurone X

gap between neurone X and neurone Y

neurone Y

electrical impulses to muscle

(a) State the name given to the junction between two neurones.

.... **(1 mark)**

(b) Explain which neurone (X or Y) on the diagram is a motor neurone. Give a reason for your answer.

....

.... **(2 marks)**

Guided   (c) Describe how neurones X and Y communicate.

When an electrical impulse reaches the end of neurone X it causes the release

of .... into the gap between the neurones. This substance

.... across the .... and causes neurone Y to

.... **(4 marks)**

3   The diagram shows a reflex arc.

(a) Describe the pathway taken by the nerve impulse in this reflex arc.

stimulus   sensory neurone   central nervous system

effector organ – muscle in the eyelid

....

....

....

....

.... **(3 marks)**

(b) What is the stimulus in this reflex arc? Give a reason for your answer.

'Give a reason' means that you have to say something that supports your answer.

....

.... **(2 marks)**

Extended response – Cells and control

Fertilisation of a human egg cell produces a zygote, a single cell that eventually gives rise to every different type of cell in an adult human.

Describe the role of mitosis in the growth and development of a zygote into an adult human.

> You will be more successful in extended response questions if you plan your answer before you start writing. Take care, because the question mentions fertilisation but it is really about growth and specialisation. Do not be tempted to talk about sexual reproduction – that is in the next topic.
>
> Your answer should include the following:
>
> • mitosis and cell division causing growth (from embryo to adult), and its importance in repair and replacement of cells
>
> • cell differentiation to produce specialised cells
>
> • the role of stem cells in the embryo as well as in the adult.
>
> Do not forget to use appropriate scientific terminology. Here are some of the words you should include in your answer:
>
> cell cycle   replication   diploid   daughter cells   specialise   differentiate

....

....

....

....

....

....

....

....

....

....

....

....

....

....

....

.... **(6 marks)**

Had a go ☐ Nearly there ☐ Nailed it! ☐

Meiosis

1 Human gametes are haploid cells. During sexual reproduction, the gametes fuse to produce a zygote.

(a) Describe what is meant by:

(i) haploid

.... **(1 mark)**

(ii) gametes

.... **(1 mark)**

(b) State the name of the male sex cells and the female sex cells in humans.

male ....

female .... **(2 marks)**

2 A cell contains 20 chromosomes. It divides by meiosis.

(a) State the number of chromosomes in each daughter cell.

.... **(1 mark)**

(b) Explain why the daughter cells are not genetically identical.

....

.... **(2 marks)**

3 The diagram below shows a cell with two pairs of chromosomes undergoing meiosis.

parent
cell

(a) State the name of the process indicated by letter **A** in the diagram.

.... **(1 mark)**

(b) Complete the diagram above to show how daughter cells are formed. **(3 marks)**

> Use the drawing as a guide. Make sure that you draw the chromosomes as they are shown, paying attention to the relative sizes.

4 Describe the importance of the two types of cell division, mitosis and meiosis.

Guided

Mitosis maintains the ...., and produces cells that are

....to the parent cell. It is used for ....

Meiosis creates .... that have .... the

number of .... Fertilisation restores the

.... **(5 marks)**

DNA

1 Our chromosomes contain genetic information. This information is held in our DNA.

(a) State the name used to describe all the DNA of an organism.

.... **(1 mark)**

Guided

(b) Describe the difference between chromosomes, genes and DNA.

This question is best answered by thinking of the definition of each of these terms.

A chromosome consists of a long molecule of ....

....

.... **(3 marks)**

2 (a) What name is given to the shape of a DNA molecule?

.... **(1 mark)**

(b) The DNA molecule is made up of a series of bases.

(i) State the number of different bases present in DNA.

.... **(1 mark)**

(ii) Describe how the two strands of the DNA molecule are linked together.

....

.... **(1 mark)**

3 The diagram shows a section of DNA.

(a) DNA is a polymer. Give **one** piece of evidence from the diagram that DNA is a polymer.

....

....

(1 mark)

(b) Identify the components A, B and C of the DNA structure. Write your answers on the diagram. **(3 marks)**

A....................................

B....................................

C....................................

You will not be expected to draw this structure from memory, but you may be expected to label the parts shown.

4 The sequence of bases on one strand of DNA is ATGGTC. What is the order of the complementary bases on the other strand?

☐ **A** TACCAG   ☐ **C** GCAACT

☐ **B** CATTAG   ☐ **D** CTGGTA   **(1 mark)**

Genetic terms

1   Draw a line to connect each statement with the corresponding genetic term.

| chromosome |

Gametes fuse to form this.

| dominant |

A short piece of DNA that codes for a characteristic.

| gene |

There are two copies of each of these in body cells.

| recessive |

The effect of this allele will only show when two copies are present in the genotype.

| zygote |

(4 marks)

2   Eye colour in humans can be controlled by two alleles of the eye colour gene. One recessive allele (b) codes for blue and one dominant allele (B) codes for brown.

> You need to know what recessive, genotype, phenotype, homozygous and heterozygous mean.

(a) (i)   State what is meant by alleles.

....   **(1 mark)**

(ii) Using eye colour as an example, explain the difference between the terms **genotype** and **phenotype**.

....

....

....

....   **(2 marks)**

(b) State the following genotypes for eye colour:

homozygous blue: ....

homozygous brown: ....

heterozygous: ....   **(3 marks)**

(c) A girl has blue eyes. Explain what her genotype must be.

....

....   **(2 marks)**

Monohybrid inheritance

1 Two plants both have the genotype Tt. The two plants are bred together.

The allele that makes the plants grow tall is represented by T, and the allele that makes plants shorter is represented by t.

> **Maths skills** Percentage probabilities from Punnett squares will always be 0, 25%, 50%, 75% or 100%, depending on the number of squares with a particular genotype (0, 1, 2, 3 or 4 squares). In fractions, probabilities will always be 0, $\frac{1}{4}$, $\frac{1}{2}$, $\frac{3}{4}$ or 1.

(a) Complete the Punnett square to give the gametes of the parents and the genotypes of the offspring.

Gametes of parent 1

Gametes of parent 2

> Take great care to complete the square correctly and use the right letters.

(2 marks)

> Guided

(b) State and explain the percentage of the offspring from this cross that will be short.

25% of the offspring from this cross will be short. I know this because

....

.... **(3 marks)**

(c) Determine the probability of the offspring from this cross being tall. Express your answer as a fraction.

....

.... **(1 mark)**

2 Fur colour in mice is controlled by two alleles, G and g. Two mice with different fur colour produced a total of 40 offspring.

(a) Complete the Punnett square for this cross.

		Parent genotype Gg	
	Parent gametes		
Parent genotype gg			

(2 marks)

(b) Homozygous recessive mice have white fur. Predict the expected number of offspring with white fur.

> Recessive alleles are shown with lowercase letters.

.... **(1 mark)**

23

Family pedigrees

1 Two healthy parents have a child who has sickle-cell anaemia, a condition caused by a recessive allele. Which **one** of the following is true?

> Questions like this can be tricky! Some answers might be true in general, but not in this particular case. You need to pick the one that is true **and** applies to this example.

☐ **A** Both parents are homozygous for the sickle-cell allele.

☐ **C** Both parents are heterozygous for the sickle-cell allele.

☐ **B** One parent is homozygous for the sickle-cell allele and the other is homozygous for the normal allele.

☐ **D** One parent is heterozygous for the sickle-cell allele and the other is homozygous for the normal allele.

(1 mark)

2 This family pedigree shows the inheritance of cystic fibrosis (CF).

CF is a genetic condition in humans caused by a recessive allele.

(a) State how many cystic fibrosis alleles an individual must inherit in order to show the symptoms of CF.

....

(1 mark)

(b) State how many males in the family pedigree have a homozygous recessive genotype.

....

(1 mark)

☐ healthy male

◯ healthy female

■ male with CF

● female with CF

Guided (c) State and explain the genotype of person 4. Use F for the normal allele and f for the recessive allele.

Person 4 does not have cystic fibrosis. This means that she must have one

.... *allele from her father. But she must*

have inherited a .... *allele from her mother.*

This means that her genotype is .... **(3 marks)**

(d) Explain the evidence that cystic fibrosis is caused by a recessive allele.

> Remember that some people are carriers of an allele that causes a genetic condition. Parents who are carriers do not have the condition, but they can pass it on to their children.

....

....

.... **(2 marks)**

Sex determination

1 (a) A baby girl is born. Explain which sex chromosome was in the sperm that fertilised the egg.

....

.... **(2 marks)**

Guided

(b) (i) Complete the Punnett square to show the sex chromosomes of both parents and all possible children.

> This is a Punnett square but you could also use a genetic diagram to show how X and Y chromosomes combine.

Father

Mother

X

X

(2 marks)

(ii) State the sex of the child in the shaded box.

.... **(1 mark)**

2 (a) A couple who have a girl wish to have a second child. Explain the chance of the couple's second child being a boy.

....

....

....

....

.... **(3 marks)**

(b) Read this statement:

> If a couple have had children and they are all girls, then the next child is more likely to be a boy.

Discuss whether you think this statement is correct.

....

....

....

.... **(2 marks)**

Variation and mutation

1 What are the causes of differences between the following?

(a) the masses of students in a year 7 class

Students in a year 7 class will show differences in mass caused by ....

variation as well as .... variation.   **(2 marks)**

(b) a pair of identical twins

Identical twins will show differences caused only by .... variation.   **(1 mark)**

2 Mr and Mrs Davies have six children. The table shows the heights of each of the six children when they reached adulthood.

Child	George	Arthur	Stanley	James	Josh	Peter
Adult height (cm)	181	184	178	190	193	179

(a) Calculate the mean height of the six Davies children. Show your working out. Give your answer to 1 decimal place.

mean height = ... cm **(2 marks)**

(b) Mr Davies is 192 cm tall and Mrs Davies is 165 cm tall. Mr Davies wonders why his children show a range of different heights. Mrs Davies wonders why the mean height of the children is not the same as the mean of her height and her husband's height. Suggest an explanation that will answer their questions.

> Don't forget to cover both genetic and at least one environmental factor. Make sure that you use scientific language such as alleles and inheritance in your answer.

....

....

....

....

.... **(4 marks)**

3 (a) Describe what is meant by a mutation.

....

....

.... **(2 marks)**

(b) State the possible effects of a mutation on the phenotype of an organism.

....

....

.... **(2 marks)**

The Human Genome Project

1 (a) State what is meant by the human genome.

....

.... **(1 mark)**

Guided

(b) State **two** advantages and **two** disadvantages of decoding the human genome.

advantage 1

A person at risk from a genetic condition will be ....

advantage 2

....

disadvantage 1

....

disadvantage 2

.... **(4 marks)**

2 Scientists have discovered that a mutation in the human *BRCA1* gene increases a woman's
risk of developing breast cancer. Give **two** benefits and **two** drawbacks to a woman of
knowing that she has this mutation.

> You need to apply your knowledge and understanding to answer this question.
> In this case, think about how it may help a woman to know that she has the
> harmful mutation in the *BRCA1* gene, and why this knowledge may cause her
> problems or concerns.

....

....

....

....

....

....

....

.... **(4 marks)**

Extended response – Genetics

Gregor Mendel carried out his work before anything was known about DNA, genes or chromosomes. He summarised his work in three laws of inheritance:

1. Each gamete receives only one factor for a characteristic.

2. The version of a factor that a gamete receives is random and does not depend on the other factors in the gamete.

3. Some versions of a factor are more powerful than others and always have an effect in the offspring.

Describe how modern knowledge of genetics has confirmed that Mendel was correct in his conclusions.

> You will be more successful in extended writing questions if you plan your answer before you start writing.
>
> Modern genetics uses terms such as genes, chromosomes, alleles, dominant and recessive, homozygous and heterozygous. Think about how Mendel's ideas and terms can be matched to these modern ones.

....

....

....

....

....

....

....

....

....

....

....

....

....

....

....

....

....

.... **(6 marks)**

Evolution

1 Darwin proposed a series of stages in evolution, including genetic variation and environmental change.

  (a) Describe what is meant by:

    (i) genetic variation

    .... **(1 mark)**

    (ii) environmental change

    .... **(1 mark)**

  (b) Explain why natural selection requires both genetic variation and environmental change.

    ....

    ....

    .... **(2 marks)**

2 Explain why, when an environment changes, some organisms within a species survive whereas others die.

> You should use scientific terms such as variation and survival in your answer.

    ....

    ....

    .... **(2 marks)**

3 When a new species is discovered, a scientist may take some of its DNA to analyse. Explain how this would help establish if this is a new species.

> **Guided**

  It will help .... the new species and to find out

  which other ....

    .... **(2 marks)**

4 It is important to complete a course of antibiotics.

Explain how stopping a course of antibiotics early can cause antibiotic resistance in bacteria.

> Darwin's theory was about natural selection and the survival of the fittest, so you should relate these to antibiotic resistance in bacteria.

    ....

    ....

    ....

    ....

    .... **(4 marks)**

Human evolution

1   Apart from the differences in body hair, using the diagrams of Ardi and Lucy state three differences between them.

    1. ....

    2. ....

    3. ....

                  (3 marks)

Ardi   Lucy

2   Some evidence for human evolution has come from the fossil record of the skull. The table below shows some of this evidence.

> You do not need to remember details such as brain sizes but you do need to remember the names and the general trends.

Name of species	Age of typical fossil (millions of years)	Brain volume (cm^3)
Ardipithecus ramidus (Ardi)	4.4	350
Australopithecus afarensis (Lucy)	3.2	400
Homo habilis	2.4	550
Homo erectus	1.8	850

(a)   Describe the relationship between when each species first appeared and brain volume.

....

.... **(2 marks)**

Guided

(b)   The first stone tools are dated from about 2.4 million years ago. Using the table, deduce what may have enabled the use of stone tools.

an increase in ....

.... **(2 marks)**

3   The diagram shows two images of stone tools.

    (a)   Explain how scientists work out the ages of stone tools.

A   B

....

....

....

.... **(2 marks)**

    (b)   Using the diagram, explain how stone tool A was held. Give reasons for your answer.

....

....

.... **(3 marks)**

Classification

1   The diagram shows the classification of some types of mammal.

GOAT   COW   LEMUR   HUMAN   GORILLA   MOUSE

Human and gorilla are the most closely related mammals shown.

Explain which pair of mammals are least closely related.

....

.... **(2 marks)**

2   Give **two** reasons why animals and plants are placed in separate kingdoms.

Plants ....

but animals ....

Plant cells have .... but .... **(2 marks)**

3   The table shows how some organisms are classified.

Classification group	Humans	Wolf	Panther
kingdom	Animalia	Animalia	Animalia
phylum	Chordata	Chordata	Chordata
class	Mammalia	Mammalia	Mammalia
order	Primate	Carnivora	Carnivora
family	Hominidae	Canidae	Felidae
genus	Homo	Canis	Panthera
species	Sapiens	Lupus	Pardus
binomial name	*Homo sapiens*	*Canis lupus*	*Panthera pardus*

Explain which two organisms in the table are most closely related.

....

.... **(2 marks)**

4   A classification system containing three domains has been suggested to replace the system containing five domains.

(a) Which domains are found in the three-domain system?

☐ **A** Plants, Animals, Prokaryotes   ☐ **C** Archaea, Eukaryota, Eubacteria

☐ **B** Prokaryotes, Eubacteria, Eukaryota   ☐ **D** Protists, Prokaryotes, Eukaryota.

(1 mark)

(b) Give the type of research that has led to the suggestion of a three-domain system.

Remember that the technology to carry out this sort of research did not exist until relatively recently.

.... **(1 mark)**

Selective breeding

1 (a) Describe what is meant by selective breeding.

....

....

....

.... **(2 marks)**

   (b) Explain how pig breeders could use selective breeding to produce lean pigs with less body fat.

> The principles of selective breeding are the same, even if you aren't familiar with this example.

....

....

....

....

.... **(3 marks)**

2 Food production can be increased by conventional plant breeding programmes.

   (a) State **three** different characteristics that could be selected for in a crop suitable for use in any country.

....

....

....

....

.... **(3 marks)**

   (b) State **two** other characteristics that might be selected for in a crop to be grown in a hot, dry part of Africa.

....

....

.... **(2 marks)**

3 Give **three** risks of selective breeding.

> **Guided**

   1. Alleles that might be useful in the future ....

   2. ....

....

   3. ....

.... **(3 marks)**

Genetic engineering

1 (a) Some potato plants have been genetically engineered so they can resist attack by insect pests. Their cells contain a gene from a different organism that produces a toxic protein. What has genetic engineering done to these potato plants?

> The phenotype of an organism is all its observable characteristics.

- ☐ **A** made no changes to their genome or phenotype
- ☐ **B** changed their phenotype but not their genome
- ☐ **C** changed their genome but not their phenotype
- ☐ **D** changed their genome and their phenotype   **(1 mark)**

(b) Resistance to insect attack is one example of a useful new characteristic given to GM crop plants.

    (i) Give **one** other example of a useful new characteristic that can be given to GM crop plants.

> You do not need to name the species of crop plant involved.

    .... **(1 mark)**

    (ii) Give **one** advantage to a farmer of planting insect-resistant potato plants.

    .... **(1 mark)**

    (iii) Suggest **one** way in which the environment may be harmed by insect-resistant potato plants.

> Think about species, other than the insect pests, that may be present in or near the potato fields.

    .... **(1 mark)**

2 Scientists have produced genetically modified mice that glow green in blue light. These 'glow mice' contain a gene naturally found in jellyfish. Describe how this genetically modified organism is

> Guided

produced.

The gene from a .... is cut out using ....

This gene is transferred to a .... embryo cell, and inserted into

a chromosome. The embryo is then allowed to develop as normal.   **(3 marks)**

3 People with Type 1 diabetes cannot produce insulin and need to inject themselves with this hormone. Until recently, insulin extracted from the pancreas of pigs was used. More recently, human insulin produced from GM bacteria has been used. Explain the advantages of using GM bacteria to produce the insulin for treating people with Type 1 diabetes.

....

....

....

.... **(4 marks)**

Extended response – Genetic modification

Humans have improved food plants over thousands of years. Compared to wild varieties, modern crop plants usually produce larger amounts of better quality food.

Discuss how selective breeding, genetic engineering and tissue culture may be used to improve crop plants. In your answer, you do **not** need to describe how these processes are carried out.

> You will be more successful in extended response questions if you plan your answer before you start writing.
>
> It may help to think about:
>
> - which processes change plants and which do not
>
> - one or more examples of desirable changes, other than the two given to you
>
> - reasons why tissue culture might be used.

....

....

....

....

....

....

....

....

....

....

....

....

....

....

....

....

....

....

....

.... **(6 marks)**

Health and disease

1 According to the World Health Organization (WHO), good health is a state of 'complete physical, social and mental well-being'. State what is meant by the following terms.

(a) physical well-being

.... **(1 mark)**

(b) mental well-being

.... **(1 mark)**

(c) social well-being

.... **(1 mark)**

2 (a) Complete the table by putting a tick in the appropriate box to show whether the disease is communicable or non-communicable.

Disease	Communicable	Non-communicable
influenza ('flu')	✔	
lung cancer		
coronary heart disease		
tuberculosis		
Chlamydia (a type of STI)		

(3 marks)

(b) Explain why you identified some diseases as communicable and others as non-communicable.

....

.... **(2 marks)**

3 HIV is a virus that can infect humans. HIV makes it easier for other pathogens to infect the human body. Suggest an explanation for how HIV does this.

┌───┐
│ Think about what types of cells are infected by the HIV virus.   │
└───┘

....

.... **(2 marks)**

4 (a) Explain how viruses cause disease.

....

....

.... **(3 marks)**

(b) Describe **two** ways in which bacteria make us feel ill.

....

....

.... **(2 marks)**

35

Common infections

1   The table shows the percentage of 15 to 49 year olds with HIV in some African countries.

African country	Percentage of 15 to 49 year olds with HIV in some African countries			
	2006	2007	2008	2009
Namibia	15.0	14.3	13.7	13.1
South Africa	18.1	18.0	17.9	17.8
Zambia	13.8	13.7	13.6	13.5
Zimbabwe	17.2	16.1	15.1	14.3

(a) Identify the country with the largest decrease in the percentage of HIV between 2008 and 2009. Show your working.

> **Maths skills** First work out what the decrease was for each country. For example, Zambia went from 13.6% to 13.5%. If you are not sure – use your calculator!

country with largest decrease .... **(2 marks)**

(b) The data for each African country follows the same overall trend. Use the data in the table to describe this trend.

....

.... **(2 marks)**

2   (a) What kind of pathogen causes Chalara ash dieback?

> Guided

☐ **A** ~~a virus~~   ☐ **C** a protist

☐ **B** a bacterium   ☐ **D** a fungus   **(1 mark)**

(b) Describe the effects of the pathogen on the trees.

....

.... **(2 marks)**

3   The table shows several diseases, the type of pathogen that causes them and the symptoms (signs of infection). Complete the table by filling in the gaps.

Disease	Type of pathogen	Signs of infection
cholera		watery faeces
	bacterium	persistent cough – may cough up blood-speckled mucus
malaria		
HIV		mild flu-like symptoms at first

(3 marks)

4   *Helicobacter* is a pathogen that causes stomach ulcers.

(a) State the type of pathogen involved.

.... **(1 mark)**

(b) Describe the symptoms it causes in infected people.

.... **(2 marks)**

How pathogens spread

1 Which of these statements about malaria is correct?

    ☐ **A**   Malaria is caused by a mosquito that invades liver cells.

    ☐ **B**   The malaria pathogen is a mosquito.

    ☐ **C**   The malaria pathogen is a protist that is spread by a vector, the mosquito.

    ☐ **D**   The malaria pathogen is a mosquito that is spread by a vector, the protist.   **(1 mark)**

2 Complete the table to show ways in which the spread of certain pathogens can be reduced.

Disease	Pathogen	Ways to reduce or prevent its spread
Ebola haemorrhagic fever		Keep infected people isolated; wear full protective clothing while working with infected people or dead bodies
tuberculosis	bacterium	

(2 marks)

3 Cholera is a disease that can spread rapidly in disaster areas when drinking water supplies are damaged. Explain **one** way that its spread could be reduced or prevented.

....

....

....

.... **(2 marks)**

4 (a) Explain why bacterial diseases such as cholera are less common in developed countries.

> Think about how these diseases are spread and how developed countries are able to control them.

....

....

....

.... **(2 marks)**

Guided ▷   **(b)** Explain why, during the 2014–15 Ebola outbreak, health workers wore full body protection when handling dead bodies.

to prevent being infected ....

because Ebola virus is present ....

....

.... **(2 marks)**

STIs

1 State what is meant by an STI.

.... **(1 mark)**

2 Which of these statements about *Chlamydia* is correct? Put an X in the box next to your answer.

☐ **A** *Chlamydia* is an STI caused by a virus.

☐ **B** A person infected with *Chlamydia* may not realise they are infected.

☐ **C** The number of new cases of *Chlamydia* diagnosed each year is falling.

☐ **D** *Chlamydia* cannot be passed from mother to baby during birth.   **(1 mark)**

3 Complete the table.

> **Guided**

Mechanism of transmission	Precautions to reduce or prevent STI
unprotected sex with an infected partner	using condoms during sexual intercourse
	supplying intravenous drug abusers with sterile needles
infection from blood products	

(3 marks)

4 (a) Explain how screening for STIs can help to reduce transmission.

....

.... **(2 marks)**

(b) Many STIs can be treated with antibiotics. Explain why HIV cannot be treated with antibiotics.

> You will need information about treating infections to be able to answer this question.
> Review page 42 of the Revision Guide if you haven't already covered this.

....

.... **(2 marks)**

Human defences

1 (a) Describe the role of the skin in protecting the body from infection.

....

.... **(1 mark)**

(b) Describe **one** chemical defence against infection from what we eat or drink.

.... **(1 mark)**

(c) (i)   Name an enzyme, found in tears, that protects against infection.

.... **(1 mark)**

(ii) Describe how the enzyme named in part (i) protects the eyes against infection.

....

.... **(2 marks)**

2 The diagram shows a section of epithelium in a human bronchiole, one of the tubes in the lung.

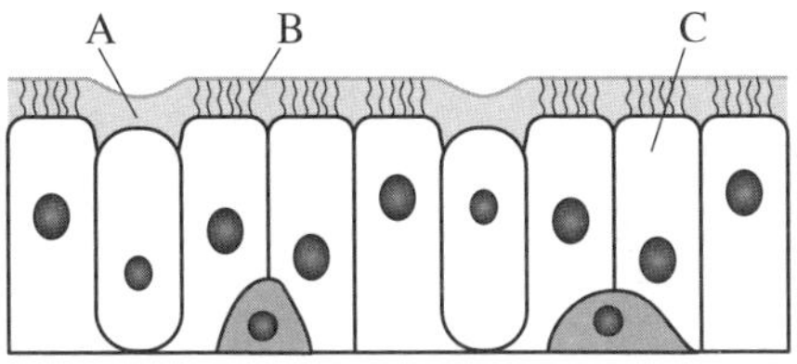

(a) (i)   State the name of the substance labelled A.

.... **(1 mark)**

(ii) Describe the role of substance A in protecting the lungs from infection.

.... **(1 mark)**

(b) (i)   State the name of the structure labelled B.

.... **(1 mark)**

▷ **Guided** ▷

(ii) Describe the part played by the type of cell labelled C in protecting the lungs from infection.

The .... on the surface of these cells move in a wave-like motion

....

....

.... **(3 marks)**

(c) Chemicals in cigarette smoke can paralyse the structures labelled B.

Explain why this increases the risk of smokers suffering from lung infections compared with non-smokers.

....

....

.... **(2 marks)**

The immune system

1   Name the type of blood cell that produces antibodies.

.... (1 mark)

2   Describe how lymphocytes help protect the body by attacking pathogens.

Pathogens have substances called .... on their surface. White blood

cells called .... are activated if they have .... that fit these

substances. These cells then ....

.... They produce large amounts of antibodies that

.... (5 marks)

3   The graph shows the concentration of antibodies in the blood of a young girl. The lines labelled A show the concentration of antibodies effective against the measles virus. The line labelled B shows the concentration of antibodies effective against the chickenpox virus.

> There is a lot to think about in this question so take it one step at a time.

(a)   At the time shown by arrow 1, there was an outbreak of measles. The girl was exposed to the measles virus for the first time in her life. Explain the shape of line A in the five weeks after arrow 1.

....

....

....

.... (4 marks)

(b)   Five months later (shown by arrow 2) there was an outbreak of measles and chickenpox. The girl was exposed to both viruses. Explain the shape of line A in the five weeks after arrow 2.

....

....

.... (3 marks)

(c)   Use lines A and B to help you answer these questions.

(i)   State whether the girl had been exposed to the chickenpox virus in the past. Explain your answer.

....

.... (2 marks)

(ii)   In the second outbreak of measles, the girl showed no symptom of measles. Explain why.

....

.... (2 marks)

Immunisation

1 Children are immunised against many childhood infections.

(a) Explain what is meant by immunisation.

....

....

....

.... **(2 marks)**

> **Guided**

(b) State what is meant by a vaccine.

A vaccine contains antigens from a pathogen, often in the form..................................

.... **(2 marks)**

(c) Explain how a vaccine prevents a person from becoming ill from a disease if they are exposed to the disease months or years after the vaccination.

....

....

....

....

....

.... **(3 marks)**

2 In 1998, a group of doctors suggested there was a connection between the MMR (measles, mumps and rubella) vaccine and autism. This made some parents afraid of having their babies vaccinated. The graph shows how the percentage of babies in the UK who were given the MMR vaccine changed over the following years.

(a) State which year had the lowest rate of vaccination.

....

(1 mark)

(b) Predict what would happen to the number of children suffering from measles in the period 1998–2004. Justify your answer.

....

....

....

.... **(2 marks)**

Treating infections

1 (a) Which of the following statements is correct?

☐ **A** An antibiotic is produced in the body to fight infection.

☐ **B** Some antibiotics are becoming resistant to bacteria.

☐ **C** Antibiotics are medicines that kill or slow down growth of bacteria in the body.

☐ **D** Antibodies are medicines that kill or slow down growth of bacteria in the body.   **(1 mark)**

(b) Explain why antibiotics can be used to treat bacterial infections in people.

....

....

....   **(2 marks)**

Guided

2 Colds are caused by viruses. A man has a very bad cold. He asks a pharmacist if an antibiotic such as penicillin would help to cure his cold.

State, with a reason, whether the pharmacist would advise the man to take penicillin.

The pharmacist's advice would be ....

The man's cold is due to a virus, so the penicillin ....

....   **(2 marks)**

3 Sinusitis causes a stuffy nose. Some patients with sinusitis were divided into two groups. One group was treated for 14 days with antibiotics whilst the other group did not receive antibiotics. Each day they were asked if they still had symptoms. The results are shown in the graph.

(graph: Percentage of patients who no longer had symptoms vs Days; curves labelled — antibiotic, ---- no antibiotic)

(a) State what you can deduce about the cause of sinusitis from the data.

> You are asked only for a deduction, not an explanation – although you might need to think about the answer to part (b) before you make your deduction!

....   **(1 mark)**

(b) Discuss whether the data supports the use of antibiotics to treat sinusitis.

> Be sure to refer to data in the graph when answering this question.

....

....

....

....   **(2 marks)**

New medicines

Guided

1 Development of a new medicine involves a series of stages. A new medicine can only move to the next stage if it has been successful in the previous stage.

(a) Complete the table to show the correct order of stages of developing a new drug.

Stage	Order
Testing in a small number of healthy people	
Discovery of possible new medicine	1
Given widely by doctors to treat patients	
Testing in cells or tissues in the lab	
Testing in a large number of people with the disease the medicine will treat	

(2 marks)

(b) (i) Describe **two** stages of preclinical testing in the development of a new medicine.

....

.... **(2 marks)**

(ii) Describe how development of a new medicine ensures that there are no dangerous side effects in humans.

.... **(1 mark)**

(c) Describe the function of a large clinical trial in developing a new drug.

> For three marks you will have to describe all of the functions; pay attention to the word 'large'.

....

....

.... **(3 marks)**

2 Scientists trialled a new medicine that was developed to lower blood pressure. They took 1000 people with normal blood pressure (group A) and 1000 people with high blood pressure (group B). Each group was divided in half; half the volunteers were given the new medicine and the other half were given a placebo (dummy medicine). At the end of the trial, the scientists measured the number of volunteers in each group who had high blood pressure.

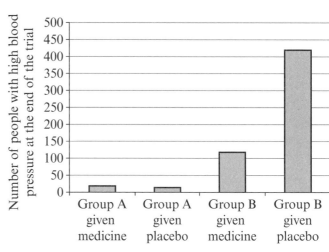

The results are shown in the bar chart.

(a) Explain why it is important for medicine trials to use large numbers of volunteers.

....

.... **(2 marks)**

(b) Use information from the bar chart to evaluate the effectiveness of this medicine.

....

.... **(2 marks)**

Non-communicable diseases

1 Explain how an infectious disease is different to a non-communicable disease.

An infectious disease is caused by a .... and is passed from ....

.... A non-communicable disease

is not passed from .... **(3 marks)**

2 State **three** factors that can affect a person's risk of developing a non-communicable disease.

1. ....

2. ....

3. .... **(3 marks)**

3 The two graphs show the prevalence of coronary heart disease (CHD) in men and women from different ethnic groups in the West Midlands. Prevalence means the percentage of people in that ethnic group who are diagnosed with the disease.

(a) State the group with the:

(i) highest incidence of CHD .... **(1 mark)**

(ii) lowest incidence of CHD.. **(1 mark)**

(b) Discuss the effect of age, sex and ethnic group on the risk of developing CHD. Use the information in the graphs in your answer.

> **Discuss** means that you need to identify the issues being assessed by the question. You need to explore the different aspects of the issue – in this case, how the incidence of CHD varies with age, sex and ethnic group.

> Make sure that you cover all three factors (age, sex and ethnic group) as well as using data from the graph to support your conclusions.

....

....

....

....

....

.... **(4 marks)**

Alcohol and smoking

1 (a) Explain how alcohol (ethanol) causes liver disease.

....

....

....

....

....

.... **(3 marks)**

(b) State why alcohol-related liver disease is described as a lifestyle disease.

.... **(1 mark)**

2 Babies whose mothers smoked while pregnant have low birth weights. Explain why.

....

....

....

.... **(2 marks)**

3 (a) State **two** diseases caused by substances in cigarette smoke.

> The question asks you to state two diseases. Remember that heart attacks and strokes are not diseases, they are the result of disease.

....

.... **(2 marks)**

Guided (b) A stroke is caused by cardiovascular disease in the brain. Explain how smoking can lead to a stroke.

Substances in cigarette smoke cause blood vessels to ....

....

....

....

.... **(3 marks)**

Malnutrition and obesity

1   The graph shows the percentage of different age groups with anaemia in a population in the USA during the 1990s.

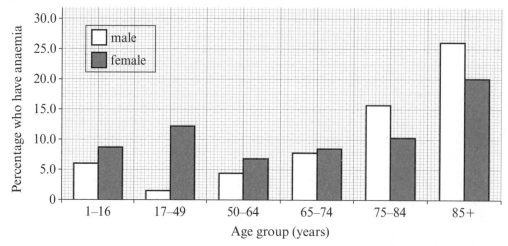

(a) Anaemia is a deficiency disease. State what is meant by deficiency disease.

.... **(1 mark)**

(b) Describe how the incidence of anaemia changes with age in males and females.

> Be sure to describe the trends in both males and females.

....

....

.... **(4 marks)**

2   The table shows the height and mass of three people.

> Guided

Subject	Mass (kg)	Height (m)	BMI
person A	80	1.80	$80/[(1.80)^2] = 80/3.24 = 24.7$
person B	90	1.65	
person C	95	2.00	

Complete the table by calculating the BMI for each person. **(2 marks)**

3   Measuring waist:hip ratio is better than BMI when predicting risk of cardiovascular disease.

(a) The table shows the waist and hip measurements of two men of the same age.

	Waist measurement (mm)	Hip measurement (m)	Waist:hip ratio
Man A	975	1.02	
Man B	914	1.06	

Complete the table by calculating the waist:hip ratio for each man. Give your answer to 2 decimal places. **(4 marks)**

(b) According to the World Health Organization, men with a waist:hip ratio greater than 0.90 have an increased risk of developing cardiovascular disease.

Explain which man is more likely to develop cardiovascular disease.

....

.... **(2 marks)**

Cardiovascular disease

1 (a) State **two** ways in which cardiovascular disease may be treated.

....

.... **(2 marks)**

(b) State **two** pieces of advice a doctor might give to a patient with high blood pressure to help them make lifestyle changes.

....

....

.... **(2 marks)**

(c) Explain why it is more important to prevent cardiovascular disease than to treat it.

....

....

.... **(2 marks)**

2 The table summarises some of the benefits and drawbacks of the different types of treatment for cardiovascular disease.

Type of treatment	Benefits	Drawbacks
lifestyle changes	no side effects	may take time to work
medication	easier to do than change lifestyle	can have side effects
surgery		
		risk of infection after surgery

Complete the table with benefits and risks of the different types of treatment.   **(3 marks)**

3 Angina is chest pain caused by narrowing of the coronary arteries. This can be treated using a stent. A stent is a wire frame that is inserted into the narrowed part of the artery. Angina can also be treated using heart bypass surgery. This is where the narrowed artery is bypassed using a section of artery or vein.

Guided

> Remember that the coronary arteries are in the heart and supply heart muscle. Think about the consequences if they become blocked.

Evaluate the use of surgery to treat angina.

Surgery can help prevent .... but costs more than inserting a

.... and surgery ....

However, ...

.... **(4 marks)**

Extended response – Health and disease

Compare and contrast the causes and treatment of communicable and non-communicable diseases.

> You will be more successful in extended response questions if you plan your answer before you start writing.
>
> Make sure you compare and contrast both communicable and non-communicable diseases.
>
> This means you need to describe the similarities and differences between the causes and the treatments; try to link these together.

....

....

....

....

....

....

....

....

....

....

....

....

.... **(6 marks)**

Photosynthesis

1 Explain why it is that food chains start with plants or algae.

> Think about what a food chain represents. You will need to use terms such as producer and biomass in your answer.

....

....

.... **(3 marks)**

2 (a) Complete the equation to show the reactants and products of photosynthesis.

.... + water → .... + ... **(2 marks)**

(b) Explain why photosynthesis is an endothermic reaction.

....

.... **(2 marks)**

3 A student knew that glucose, produced by photosynthesis, is converted into starch in leaves. She also knew that iodine solution turns blue–black in the presence of starch. The student carried out two experiments to investigate photosynthesis.

(a) The box shows what happened in the first experiment.

> **Experiment 1**
>
> A plant was kept in the dark for 48 hours. This removed all starch from its leaves. Some of the leaves were covered in foil. The plant was then placed on a sunny windowsill. Two leaves were tested for starch a few hours later:
> • the leaf covered in foil did not produce a blue–black colour with iodine
> • the leaf left uncovered produced a blue–black colour with iodine.

Explain what Experiment 1 shows about photosynthesis.

....

.... **(2 marks)**

(b) In the second experiment, a plant with variegated leaves was used. Variegated leaves are green with white patches, rather than completely green.

> **Experiment 2**
>
> The plant was kept in the dark for 48 hours to remove all starch from its leaves. The plant was then placed on a sunny windowsill. A leaf was tested for starch a few hours later:
> • only the green parts of the leaf produced a blue–black colour with iodine.

Explain what Experiment 2 shows about photosynthesis.

> What is present in the green parts of the leaf that is not present in the white parts?

....

.... **(2 marks)**

Had a go ☐   Nearly there ☐   Nailed it! ☐

Limiting factors

Guided

1   Describe what is meant by a limiting factor.

This is a factor or variable that stops the rate of something ....

The rate will only increase if this factor is .... **(2 marks)**

2   (a)   Name **one** factor other than carbon dioxide concentration and light intensity that limits the rate of photosynthesis.

.... **(1 mark)**

(b)   Describe how you could measure the rate of photosynthesis using algal balls.

....

....

....

....

....

.... **(3 marks)**

3   The graph shows how the rate of photosynthesis changes with light intensity. The data show the rate at three different concentrations of carbon dioxide.

(a)   Describe how increasing the concentration of carbon dioxide changes the rate of photosynthesis.

.... **(1 mark)**

(b)   Commercial growers often increase the concentration of carbon dioxide in their greenhouses.

Explain how this will increase the yield of crops grown in the greenhouse.

....

....

....

.... **(2 marks)**

Guided

(c)   Explain how the rate of photosynthesis could be increased further.

You could increase the .... as this

would make photosynthesis happen... **(2 marks)**

 **Practical skills**

Light intensity

1 Some students wanted to investigate how the rate of photosynthesis in pond weed changed with light intensity. They did this by putting a lamp at different distances from some pond weed in a test tube. They counted the number of bubbles produced by the plant. Here are the data they collected.

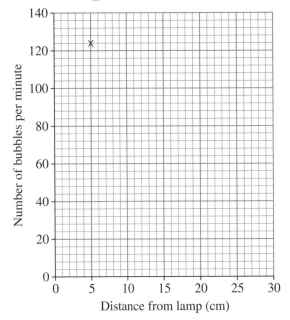

Distance from lamp (cm)	5	10	15	20	25	30
Number of bubbles per minute	124	88	64	42	28	16

(a) Complete the graph to show the results in the table. **(2 marks)**

> Mark the points accurately on the grid (to within half a square) using the table of data. Then draw a line of best fit through these points. This line does not have to be straight.

(b) Use your graph to find the number of bubbles you would expect in 1 minute if the lamp was placed 12 cm from the pond weed.

.... **(1 mark)**

(c) Describe how the rate of bubbling changes as the distance from the lamp increases.

....

.... **(2 marks)**

(d) (i) State **one** safety step you should take.

....

.... **(1 mark)**

(ii) Explain **one** step you should take to ensure that your results are reliable.

> Explain means that you have to say what the step is and why you need to take that step.

....

....

.... **(2 marks)**

⟩ **Guided** ⟩ (e) Describe how you could use a light meter to improve the experiment.

You could use the light meter to measure the .... at each

distance and then plot a graph of ....

.... **(2 marks)**

Specialised plant cells

1 The diagram shows part of a plant tissue specialised for transport.

(a) State the name of this type of tissue.

.... **(1 mark)**

A —

B —

mitochondrion

vacuole

companion cell

sieve cell

> **Guided** (b) Explain how the features labelled A and B are adapted to the function of this tissue.

A ....

....

B There is a small amount of cytoplasm so ....

.... **(4 marks)**

(c) Explain why companion cells have many mitochondria.

> Mitochondria supply energy. You need to give this information AND explain why companion cells need lots of energy.

....

....

....

.... **(2 marks)**

2 (a) State the name of the vessels used to transport water in plants.

.... **(1 mark)**

> **Guided** (b) Describe **three** ways in which these vessels are adapted for their function.

1. The walls are strengthened with lignin rings to...

....

2. ....

....

3. ....

.... **(3 marks)**

Transpiration

1 A student set up the following experiment to investigate transpiration.

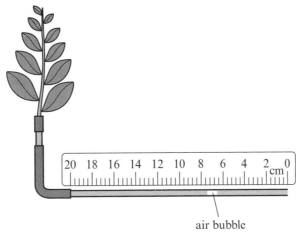

air bubble

(a) State what is meant by the term **transpiration**.

....

.... **(2 marks)**

(b) State which part of the plant regulates the rate of transpiration.

.... **(1 mark)**

(c) For each of the following situations, predict what will happen to the air bubble. Give a reason for each answer.

> The fan simulates a windy day.

(i) A fan is started in front of the plant.

....

.... **(2 marks)**

(ii) The undersides of the leaves of the plant are covered with grease.

....

.... **(2 marks)**

2 (a) Explain how the guard cells open and close.

....

....

....

.... **(3 marks)**

Guided

(b) The stomata are open during the day but closed at night. Explain why, in very hot weather, plants wilt during the day but recover during the night.

The stomata are open during the day, so water is lost by ....

faster than it can be absorbed by the .... Water is lost from the vacuoles

and the plant wilts. At night, the stomata ....

.... **(3 marks)**

53

Translocation

1 (a) State what is meant by **translocation**.

....

.... **(1 mark)**

(b) What is the name of the plant tissue responsible for translocation?

☐ **A** phloem

☐ **B** xylem

☐ **C** meristem

☐ **D** mesophyll   **(1 mark)**

2 (a) Describe how radioactive carbon dioxide can be used to show how sucrose is transported from a leaf to a storage organ such as a potato.

Guided

Radioactive carbon dioxide is supplied to the leaf of a plant. ....

....

....

....

.... **(3 marks)**

(b) An inhibitor is a substance that can stop an enzyme or process working. Predict the effect on translocation of adding an inhibitor of active transport to the leaf. Give a reason for your answer.

What type of transport is involved in translocation?

....

....

.... **(2 marks)**

3 The table lists some of the structures and mechanisms involved in movement of water and sucrose in the plant. Put an X in each row of the table to show whether the structure or mechanism is involved in the transport of water or the transport of sucrose.

You might need to revise transpiration on page 53 of the Revision Guide before answering this question.

Structure or mechanism	Transport of water	Transport of sucrose
xylem		
phloem		
pulled by evaporation from the leaf		
requires energy		
transported up and down the plant		

(5 marks)

Water uptake in plants

1   The rate of transpiration increases if the light intensity or temperature is increased. Complete the table by placing a tick (✓) against information that could be used to explain this.

Information	Increased light intensity	Increased temperature
stomata become more open		
stomata become more closed		
water molecules have less energy		
water molecules have more energy		
rate of evaporation increased	✓	✓
rate of evaporation decreased		

(3 marks)

2   Some students investigated the rate at which water evaporated from leaves using this apparatus.

The students measured how far the air bubble travelled up the capillary tube in 5 minutes with the fan on, and with the fan off. They found that the bubble moved 90 mm with the fan off and 130 mm with the fan on.

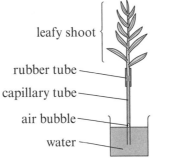

leafy shoot

rubber tube

capillary tube

air bubble

water

(a) Explain the results the students collected.

....

....

....

.... **(3 marks)**

(b) The speed of the fan was increased, and it was found that the rate that the bubble moved did not increase. Explain, in terms of the plant's response, why this was the case.

> In this question, you need to think about what the plant does as the speed of the wind increases.

....

....

....

.... **(3 marks)**

(c) The capillary tube had a diameter of 0.5 mm. Calculate the rate of transpiration in mm³/min when the fan was off.

volume of tube $= \pi \times (\frac{diameter}{2})^2 \times$ length

$= \pi \times ($.............$)^2 \times$ .... $=$ .... mm³

rate $= \dfrac{\text{volume}}{\text{time}} = \dfrac{.........................}{.........................}$ mm³/min

$=$ .... mm³/min (rounded to 1 decimal place)   **(2 marks)**

Extended response – Plant structures and functions

 Plant stomata are closed at night and open during the day. Explain how this process occurs, and how it affects the movement of water and mineral ions through the plant.

> You will be more successful in extended response questions if you plan your answer before you start writing.
>
> This question is about water loss, so make sure you explain:
>
> - how stomata open and close
> - the role of stomata in controlling water loss
> - the role of transpiration in providing mineral ions needed for growth.

....

....

....

....

....

....

....

....

....

....

....

....

.... **(6 marks)**

Hormones

Guided

1 (a) Describe how hormones behave like 'chemical messengers'.

Hormones are produced by .... and are released

into the .... They travel round the body until they reach

.... which responds by releasing

.... **(4 marks)**

(b) Describe **two** ways in which hormones and nerves communicate differently.

> Make sure you describe two ways and that they are differences, not similarities.

....

.... **(2 marks)**

2 The diagram shows the location of some endocrine glands in the body. Write in the name of each gland on the corresponding label line.

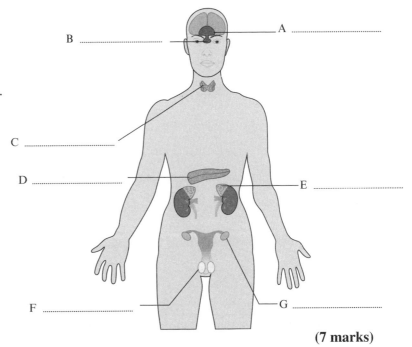

(7 marks)

3 Complete the table showing where some hormones are produced and where they have their action.

Hormone	Produced in	Site of action
thyroxine	thyroid gland	various organs, including the heart
FSH and LH		ovaries
insulin and glucagon		liver, muscle and adipose (fatty) tissue
adrenalin		various organs, e.g. heart, liver, skin
progesterone		uterus
testosterone		male reproductive organs

(5 marks)

The menstrual cycle

1 State **two** of the hormones that control the menstrual cycle.

.... **(2 marks)**

2 The diagram below shows the timing of some features in a menstrual cycle.

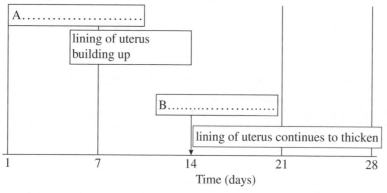

A......................

lining of uterus building up

B......................

lining of uterus continues to thicken

1   7   14   21   28

Time (days)

(a) Fill in the two missing labels, A and B, on the diagram. **(2 marks)**

(b) Mark with an X on the diagram the point at which fertilisation is most likely to occur. **(1 mark)**

(c) Describe what happens during days 1–5 of the cycle.

....

....

.... **(2 marks)**

3 (a) Explain how hormonal contraception prevents pregnancy.

> **Guided**

Pills, implants or injections release hormones that prevent ....

and thicken ....

preventing .... from passing. **(3 marks)**

(b) The table shows the success rates of different methods of contraception.

Method of contraception	Success rate (% of pregnancies prevented)
hormonal pill or implant	> 99%
male condom	98%
diaphragm or cap	92–96%

(i) Explain why the actual success rate can sometimes be lower than the figures shown.

....

.... **(2 marks)**

(ii) Some methods of contraception protect against STIs as well as reducing the chance of pregnancy. Complete the table by placing a tick (✓) against the correct features of each method.

Method of contraception	Reduces chance of pregnancy (✓)	Protects against STIs (✓)
hormonal pill or implant		
male condom		
diaphragm or cap		

(3 marks)

Blood glucose regulation

1 The table shows the events that happen after a person eats a meal. Complete the table to show the order in which the events take place.

Event	Order
Pancreas increases secretion of insulin.	
Blood glucose concentration falls.	
Blood glucose concentration rises.	1
Insulin causes muscle and liver cells to remove glucose from blood and store it as glycogen.	
Pancreas detects rise in blood glucose concentration.	

(3 marks)

2 The graph shows the effect of a hormone on the concentration of glucose in the blood.

(a) Calculate the percentage change in blood glucose concentration after eating dinner.   **(2 marks)**

Remember that you can draw lines on the graph if this helps you to work out your answer.

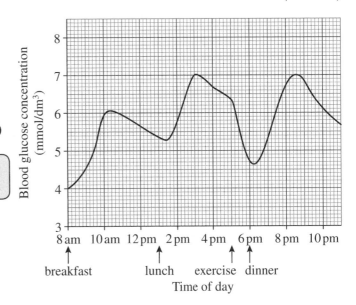

(b) Explain why there is a decrease in blood glucose concentration between 10 am and 1 pm.

....

.... **(2 marks)**

(c) Explain why the blood glucose concentration decreases more rapidly between 5 pm and 6 pm.

....

....

.... **(2 marks)**

Diabetes

1 (a) The graph shows the percentage of people in one area of the USA in the year 2000 who have Type 2 diabetes, divided into groups according to body mass index.

Describe how the percentage of people with Type 2 diabetes changes as the BMI increases.

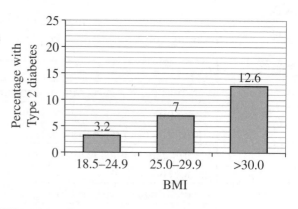

....

.... **(2 marks)**

(b) Two 45-year-old males from the area of the USA studied in part (a) wanted to estimate their chances of developing Type 2 diabetes.

(i) George was 180 cm tall and had a mass of 88 kg. Calculate his BMI and use this to evaluate his risk of developing Type 2 diabetes.

$$BMI = \frac{weight\ (kg)}{(height\ (m))^2}$$

.... **(3 marks)**

(ii) Donald had a waist measurement of 104 cm and a hip measurement of 102 cm. The World Health Organization classes a waist : hip ratio of >0.9 as obese. State and explain whether Donald has an increased risk of developing Type 2 diabetes.

....

....

.... **(3 marks)**

2 (a) Explain how helping people to control their diets might help to reduce the percentage of people in the population who have diabetes.

> **Guided**

Controlling diets will help to...

Fewer obese people means.. **(2 marks)**

(b) (i) Explain why people with Type 1 diabetes are treated with insulin but most people with Type 2 diabetes are not.

....

.... **(2 marks)**

(ii) Explain why a person with Type 1 diabetes will sometimes wait to see how large a meal is before deciding how much insulin to inject.

....

.... **(2 marks)**

Extended response – Control and coordination

Compare how Type 1 and Type 2 diabetes are caused and how they are treated.

> You will be more successful in extended response questions if you plan your answer before you start writing.
>
> Make sure that you cover the causes of each type of diabetes and link this to the type of treatment.

....

....

....

....

....

....

....

....

....

....

....

....

....

....

....

....

....

....

....

....

....

....

.... (6 marks)

Exchanging materials

1   Substances are transported into and out of the body. Describe where and why the following substances are removed from the bloodstream.

(a) Water ....

....

.... **(2 marks)**

(b) Urea ....

....

.... **(2 marks)**

2   Humans and other mammals need to exchange gases with their environment. Describe where and why this exchange happens.

....

....

.... **(3 marks)**

3   Absorption of digested food molecules takes place in the small intestine. The small intestine has a surface adapted to assist this process.

> Guided >

(a) Describe how the small intestine is adapted to help to absorb food molecules.

The surface of the small intestine is covered with .... These help

by increasing ....

.... **(2 marks)**

(b) Explain why the structures described in part (a) have thin walls.

....

.... **(2 marks)**

4   The diagram shows a flatworm and an earthworm.

> Guided >

The two worms are similar in size. Explain why the flatworm does not have an exchange system or a transport system whereas the earthworm has a transport system (heart and blood vessels).

The flatworm is very flat and thin which means it has a large ....

....

....

....

.... **(4 marks)**

Alveoli

Guided

1 (a) Describe how gas exchange takes place in the lungs.

Oxygen diffuses from .... into ....

Carbon dioxide diffuses from .... into .... **(2 marks)**

(b) State and explain **two** ways in which the structure of the alveoli is adapted for efficient gas exchange.

Millions of alveoli create a large ....

for the .... of gases. Each alveolus is closely associated with

a .... Their walls are one ....

.... **(4 marks)**

2 Explain the importance of continual breathing and blood flow for gas exchange.

....

....

....

.... **(2 marks)**

3 Emphysema is a type of lung disease where elastic tissue in the alveoli breaks down. The figure shows the appearance of an alveolus damaged by lung disease compared with a healthy alveolus.

Healthy alveolus   Alveolus damaged by lung disease

Explain how emphysema affects the person.

Think about what effects the changes seen in emphysema would have on gas exchange and how this would then affect the person.

....

....

....

....

....

.... **(3 marks)**

Blood

1 Draw **one** line from each blood component to a correct function.

Blood component	Function
plasma	carries other blood components
platelet	part of the body's immune system
red blood cell	involved in forming blood clots
white blood cell	carries oxygen

(3 marks)

2 Blood contains red blood cells.

(a) Name the cell structure, normally found in cells, that is missing in human red blood cells.

.... **(1 mark)**

(b) Name the compound in red blood cells that gives them their colour.

.... **(1 mark)**

> **Guided**

(c) The diagram shows some red blood cells.

Describe **two** ways in which red blood cells are adapted to carry out their function.

Their biconcave shape gives them a large ....

for diffusion to happen efficiently. They are also flexible, which lets them ....

.... **(2 marks)**

3 The plasma transports soluble products of digestion, including glucose and amino acids. Name **two** waste substances transported by the plasma.

1 ....

2 .... **(2 marks)**

4 Explain how platelets help to protect the body from infection.

....

....

.... **(3 marks)**

5 White blood cells usually make up about 1% of the blood and include lymphocytes and phagocytes.

(a) Explain why the number of lymphocytes increases during infection.

....

....

.... **(3 marks)**

(b) Describe how phagocytes help protect the body.

....

.... **(2 marks)**

Blood vessels

Guided

1 (a) Describe the structure of an artery.

An artery has .... walls. These walls are composed of two types of fibres:

.... tissue and .... fibres.   **(3 marks)**

(b) Explain how the structure of the artery wall makes blood flow more smoothly in arteries.

....

....

....   **(2 marks)**

2 Blood needs to penetrate every organ in the body. This is made possible by capillaries.

(a) Describe how the capillaries are adapted for this function.

> Make sure that you describe here and save the explanation for part (b).

....

....

....   **(2 marks)**

(b) Explain how the features you have described are important for the function of capillaries.

....

....

....   **(2 marks)**

3 (a) Veins carry blood away from the organs of the body to the heart.

(i) Explain why there is a difference in the thickness of the walls of arteries and veins.

....

....

....   **(2 marks)**

(ii) Explain how muscles and valves work together to help return blood to the heart.

....

....

....   **(2 marks)**

(b) Explain why a nurse taking blood from a patient will insert a needle into a vein rather than an artery.

> There are two possible answers here – you need to consider either the structure of the different blood vessels, or else the way in which each transports the blood they contain.

....

....

....   **(2 marks)**

The heart

1 The heart is connected to four major blood vessels. Describe where each vessel carries blood.

Blood vessel	Carries blood:	
	from	**to**
aorta	heart	body
pulmonary artery		
pulmonary vein		
vena cava		

(4 marks)

2 (a) Explain why the heart consists mostly of muscle.

....

.... **(2 marks)**

(b) Describe the route taken by blood through the heart from the vena cava to the aorta.

....

....

.... **(3 marks)**

3 The diagram shows a section through the human heart.

Remember that the heart is drawn and labelled as if you were looking at the heart in someone's body. So the right side of the heart is actually on the left side of the page!

(a) State the name of the part of the heart labelled A and describe its function.

....

.... **(2 marks)**

(b) Explain the function of the part labelled B.

....

.... **(2 marks)**

(c) Explain why the muscle at C needs to be thicker than on the other side of the heart.

....

....

.... **(3 marks)**

Aerobic respiration

1   Read the following passage and answer the questions that follow.

> Aerobic respiration happens in muscle cells in the body. The muscle cells are surrounded by blood vessels. The substances needed for respiration are transferred to the muscle cells by diffusion, and the waste products are removed.

   (a) Name the substances needed for respiration in muscle cells.

   .... and .... **(2 marks)**

> Guided

   (b) State the meaning of the term **diffusion**.

   Diffusion is the movement of substances from .... to .... concentration.

   (1 mark)

2   (a) State the location in the cell where most of the reactions of aerobic respiration occur.

   ....
   (1 mark)

   (b) Explain how cellular respiration helps maintain the body temperature.

   ....

   ....

   .... **(2 marks)**

   (c) State **one** way that animals use energy from respiration, other than to maintain their body temperature.

   ....

   .... **(1 mark)**

3   The blood supplies cells with the substances needed for aerobic respiration, as well as removing waste products.

   (a) Write a word equation for aerobic respiration.

   .... **(1 mark)**

   (b) State the name of the smallest blood vessels that carry blood to the respiring cells.

   .... **(1 mark)**

4   (a) Explain why all organisms respire continuously.

   ....

   .... **(2 marks)**

   (b) Plants can use energy from sunlight in photosynthesis. Explain why plants also need to respire continuously.

   > Photosynthesis uses light energy in production of glucose; it does not release energy that can be used in other processes. Think about why plants need energy from respiration.

   ....

   .... **(2 marks)**

Anaerobic respiration

1 Humans can respire in two ways: using oxygen (aerobic) and without using oxygen (anaerobic).

> Make sure that you understand what is produced in both aerobic and anaerobic respiration.

(a) Compare the amounts of energy transferred by aerobic and anaerobic respiration.

....

.... **(2 marks)**

(b) Describe the circumstances under which anaerobic respiration occurs.

....

.... **(2 marks)**

2 In track cycling, a 'sprint' event begins with several slow laps in which the riders try to get a tactical advantage. These slow laps are followed by a very fast sprint to the finishing line.

(a) Describe and explain how the cyclists' heart rates change during the course of the race.

....

....

.... **(3 marks)**

(b) After the race the cyclists will cycle on a stationary bicycle for 5–10 minutes. Explain why they do this.

....

.... **(2 marks)**

3 The graph shows how oxygen consumption changes before, during and after exercise. The intensity of the exercise kept increasing during the period marked 'Exercise'.

(a) Explain the shape of the graph during the period marked 'Exercise'.

....

....

.... **(3 marks)**

> **Guided**

(b) Explain the shape of the graph during the period marked 'Recovery'.

During exercise there is an increase in the concentration of ....

....

....

.... **(2 marks)**

68

 **Rate of respiration**

1   The diagram shows a respirometer used to investigate the rate of respiration in germinating peas.

| syringe containing air |
| 3-way tap |
| ruler |
| capillary tubing   blob of liquid |
| respiring peas |
| wire gauze |
| water bath |
| potassium hydroxide |

> This is one of the core practicals so you should be able to answer questions on the apparatus.

State the role of the following and give a reason for your answer:

(a) the water bath

....

....

....   **(2 marks)**

Guided

(b) the potassium hydroxide

absorbs carbon dioxide produced by the .... so that

....

....   **(2 marks)**

(c) the tap and syringe containing air

....

....

....   **(2 marks)**

2   (a) Explain how the apparatus allows you to measure the rate of respiration in the seeds.

> The movement of the liquid blob gives you information about the uptake of oxygen. You need to state this, explain how you measure the movement and then how you calculate the rate of respiration.

....

....

....   **(3 marks)**

(b) Describe how you would use the apparatus to investigate the effect of temperature on the rate of respiration in peas.

....

....

....   **(3 marks)**

Changes in heart rate

1   Cardiac output can be calculated using the equation:
    cardiac output = stroke volume × heart rate.

(a)   What is meant by the term **stroke volume**?

.... **(1 mark)**

(b)   A man has an average stroke volume of 75 cm³ and a heart rate of 60 beats/minute.

(i)   Calculate his cardiac output. Show your working out and give the correct units.

Cardiac output .... **(3 marks)**

(ii)   Explain the change in cardiac output when the man starts to exercise.

....

....

.... **(3 marks)**

2   The graph shows the pulse rate of an athlete at rest, and after 5 minutes of different types of exercise.

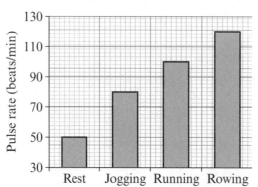

> Remember to show all your steps in the calculation.

Guided

(a)   Calculate the percentage increase in pulse rate between jogging and running.

100 beats/min – 80 beats/min = .../min

(............/80) x 100 =

Percentage increase .... **(2 marks)**

(b)   State why the pulse rate is highest when the athlete is rowing.

....

.... **(1 mark)**

(c)   The pulse is a measure of heart rate. At rest, the cardiac output of the athlete is 4000 cm³/min.
      Calculate the stroke volume, in cm³, of the athlete at rest.

Stroke volume .... cm³   **(2 marks)**

Extended response – Exchange

The diagram shows the main features of the human heart and circulatory system.

Describe the journey taken by blood around the body and through the heart, starting from when it enters the right side of the heart. In your answer, include names of major blood vessels and chambers in the heart.

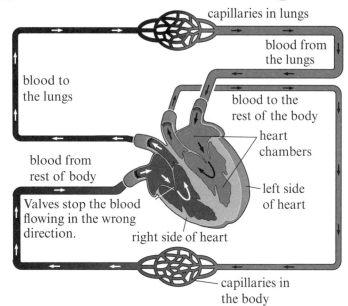

> You will be more successful with extended response questions if you plan your answer before you start writing.
>
> You do not need to explain how the different components of the heart and circulatory system work. It may help your plan if you follow the blood around the diagram with a finger, writing the name of each blood vessel or chamber in order as you go. You do not need to identify any blood vessels in the 'rest of the body' other than the aorta.

....

....

....

....

....

....

....

....

....

....

....

....

....

....

.... (6 marks)

Ecosystems and abiotic factors

1 Draw lines to connect each term with its definition.

Term		Definition
community		a single living individual
organism		all the living organisms and the non-living components in an area
population		all the populations in an area
ecosystem		all the organisms of the same species in an area

(4 marks)

Guided

2 A student surveyed the distribution of a species of lichen growing on the trunk of a tree. She used a small quadrat to measure the percentage cover by these lichens on the south- and north-facing sides of the tree.

	Light intensity (lux)			
	Reading 1	**Reading 2**	**Reading 3**	**Mean**
South side	275.5	368.1	326.8	
North side	195.7	282.1	205.1	

Percentage cover											
South side						**North side**					
1	2	3	4	5	Mean	1	2	3	4	5	Mean
48	20	28	92	8	39	4	4	4	4	6	4

(a) Complete the upper table to show the mean light intensity for each side.

south side $\dfrac{(275.5 + 368.1 + 326.8)}{3} =$

north side .... **(2 marks)**

(b) The student concluded that the lichen was better adapted to conditions on the south side. Justify her conclusions.

> In this part of the question you are asked to say only whether she was right and, if so, why. Don't try to explain her results.

....

....

.... **(2 marks)**

(c) Light intensity is an abiotic factor. Explain one other abiotic factor that might be responsible for the different distribution of lichen.

....

....

.... **(2 marks)**

Biotic factors

1   Meerkats are animals that live in packs and are found in the desert areas of southern Africa. The pack of meerkats is led by a dominant pair of meerkats, known as the alpha male and female.

> In this question, you will be asked to think about aspects of the behaviour of the meerkats. Remember to link your answer to the ideas that these animals will compete with each other for resources.

(a) State what the term **biotic factors** means.

.... **(1 mark)**

(b) Only the alpha male and alpha female breed. Suggest an explanation for why younger male meerkats will often try to fight the alpha male.

....

.... **(2 marks)**

(c) When meerkat packs become very large, they often split into smaller packs. The new pack will often move some distance from the original pack. Explain the reasons why a large pack may need to split up.

> Make sure that you know what the command words mean. **Explain** means give a reason why. **Suggest** means that you need to apply your knowledge to a new situation. **Describe** means say what is happening.

....

.... **(2 marks)**

2   The drawing shows a male peacock.

State and explain **one** adaptation, seen in the diagram, that helps the peacock attract a mate.

....

....

....

.... **(3 marks)**

3   The diagram shows a cross-section through a tropical rainforest.

> Guided

(a) Some trees are called emergent. They break through the rest of the rainforest canopy. Explain the advantage to these trees of emerging from the canopy.

The trees emerge through the canopy

to get .... for

more .... **(2 marks)**

(b) The soil in a rainforest is often poor as the minerals are washed away (leached). Suggest an explanation of how trees in the rainforest may adapt in response to a leached soil.

....

.... **(2 marks)**

Parasitism and mutualism

1   The table has statements about relationships between species. Place an X in one of the boxes on each line to show whether the statement applies to parasitism only, mutualism only, or both parasitism and mutualism.

Statement	Parasitism only	Mutualism only	Both parasitism and mutualism
There is interdependence, where the survival of one species is closely linked with another species.			
One species lives inside the intestine of another and absorbs nutrients from the digested food.			
One species lives inside another and receives food from the host. In return, the host receives nutrients.			

(3 marks)

2   Cleaner fish are small fish that feed on parasites on the skin of sharks. Describe how the cleaner fish and the sharks benefit from a mutualistic relationship.

> Guided

Cleaner fish get food by...

....

This helps the shark because ....

.... (2 marks)

> An exam question may ask you about the benefits to one organism or to both. Make sure that you read the question carefully!

3   The scabies mite is a tiny arthropod that burrows into human skin and lays its eggs. Infection by the scabies mite causes severe itching and a lumpy, red rash that can appear anywhere on the body. Explain why the scabies mite is a parasite and not a mutualist.

> This type of question is expecting you to apply your understanding of science to a situation that you may not be familiar with. You will have been taught about organisms that behave in a similar way to the scabies mite – use what you know about these organisms but apply it to the scabies mite.

....

....

....

.... (2 marks)

Fieldwork techniques

1 A gardener goes into his garden every night at 7 pm and counts the number of slugs in the same 1 m² area of his flower bed. He records his results in a table.

Day	Monday	Tuesday	Wednesday	Thursday	Friday	Saturday	Sunday
Number of slugs	11	12	7	12	8	8	12

(a) Describe how the gardener could make sure the 1 m² area of the flower bed was chosen at random on the first day.

....

.... **(2 marks)**

> Guided

(b) Why does the gardener use the same area each time?

Using the same area means that his experiment is .... **(1 mark)**

(c) Describe **one** way in which the gardener could improve the repeatability of the data that he collected.

....

.... **(2 marks)**

2 A class is investigating the number of clover plants on a football pitch. The pitch measures 100 m by 65 m. The class wants to find the total number of clover plants in the field. The teacher gives the class a 1 m × 1 m quadrat.

$$\text{mean number of plants} = \frac{\text{total number of plants in all quadrats}}{\text{number of quadrats}}$$

(a) Explain how the class can use the quadrat to estimate the mean number of clover plants in a 1 m² area.

....

.... **(2 marks)**

(b) The class finds that the mean number of clover plants in an area of 1 m × 1 m is 7. Estimate the number of clover plants on the whole football pitch.

....

....

.... **(3 marks)**

3 Describe how you would use a belt transect to investigate the distribution of broad-leaved plants growing alongside a path that started at a road, crossed a small field and entered a wood.

> Make sure that you describe use of quadrats, the measurements you would take and what you would record.

....

....

....

.... **(3 marks)**

75

Organisms and their environment

Practical skills

1   Limpets are animals that have a shell and live on rocks that are underwater some or all of the time. They can be found in the sea, or in rock pools on the beach. A scientist is investigating the distribution of limpets on the beach.

(a) Explain how the scientist could use a transect to investigate the distribution of limpets.

....

....

.... **(3 marks)**

The scientist sets up three different transects and measures the numbers of limpets on each one. His data are shown in the table.

Distance from sea (metres)	Number of limpets			Mean number of limpets
	Transect 1	Transect 2	Transect 3	
0.5	20	23	20	21
1.0	18	16	17	17
1.5	13	13	13	13
2.0	10	8	9	
2.5	5	6	4	5

(b) Calculate the mean number of limpets at 2.0 m from the sea in this investigation.

.... **(2 marks)**

(c) What conclusion can be made from his investigation?

> Your conclusion should describe how the distribution of limpets changes along the transect.

....

....

.... **(3 marks)**

Guided

2   A scientist investigated the distribution of bluebells in a large wood. She started on the edge of the wood, and measured a line going deeper into the wood. Every 2 m into the woodland, she placed a quadrat and counted the number of bluebells in the quadrat. She also measured the light intensity at each quadrat.

(a) Describe **one** way the scientist could alter her method to collect more accurate data.

Instead of placing a quadrat every 2 m, the scientist could ....

.... and use a .... quadrat than before. **(2 marks)**

(b) The scientist obtained the following data:

Distance from edge of wood in metres	0	2	4	6	8	10	12	14	16
Number of bluebells	0	7	15	22	25	21	16	10	8

Suggest an explanation for these results.

....

.... **(2 marks)**

Human effects on ecosystems

1   Fish can be farmed or caught from the wild. State **one** advantage of fish farming, and **one** disadvantage.

Advantage reduces fishing of ....

Disadvantage ....

.... **(2 marks)**

2   A non-indigenous species is not naturally found in a particular place. For example, the cane toad is a non-indigenous species in Australia that was introduced to control insect pests. State **one** other advantage of introducing a non-indigenous species, and **one** disadvantage.

Advantage ....

....

Disadvantage may reproduce rapidly as they ....

....

.... **(2 marks)**

3   The graph shows the mass of fertiliser used in the world from 1950 to 2003.

(a) Calculate the percentage increase in fertiliser use from 1950 to 2003.

> Make sure that you read the graph carefully to get the correct figures for your calculation. You will get one mark for showing the correct calculation and one mark for the correct answer.

....   **(2 marks)**

(b) Suggest an explanation for the change in the mass of fertiliser used worldwide since 1950.

....

.... **(2 marks)**

(c) Describe an environmental problem caused by over-use of fertilisers.

....

....

.... **(2 marks)**

Biodiversity

1 (a) State what is meant by **reforestation**.

....

.... **(1 mark)**

(b) Describe **two** advantages of reforestation.

....

....

....

.... **(2 marks)**

2 Explain the importance to humans of conservation.

....

....

....

.... **(2 marks)**

3 Yellowstone National Park in the USA is the natural home of many species. Deer eat young trees, stopping them from growing. The population of deer increased so much that Park authorities decided to reintroduce some wolves.

> Guided

The wolves killed some of the deer for food. The deer moved away from river areas because they were more easily hunted by the wolves there. The wolves also killed coyotes, which are predators that eat rabbits.

The reintroduction of wolves led to major improvements in the biodiversity of the Park. This included increases in the populations of rabbits, bears, hawks and other birds. It also reduced the amount of soil washed into the rivers.

Describe the ways in which the reintroduction of the wolves may have caused the biodiversity to improve.

> This is an example of having to apply knowledge you have learned in this and other units to an unfamiliar situation.

The numbers of trees will increase because ....

This means there will be more food for ....

There will be more rabbits because ....

If there are more rabbits, there will be more food for ....

More trees also mean that there will be more ....

...........................and less .... **(6 marks)**

The carbon cycle

1 Complete the diagram of the carbon cycle by writing the names of the processes in the boxes.

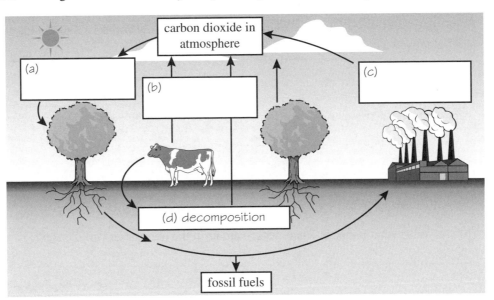

(a) _____

carbon dioxide in atmosphere

(b) _____

(c) _____

(d) decomposition

fossil fuels

(4 marks)

2 Explain why microorganisms are important in recycling carbon in the environment.

....

....

.... **(2 marks)**

> In questions about the carbon cycle, you will be expected to make links between photosynthesis, respiration and combustion, and the amount of carbon dioxide in the air.

3 (a) The diagram shows a fish tank. Explain how carbon is recycled between organisms in the fish tank.

....

....

....

....

....

....

.... **(4 marks)**

(b) Explain why it is important that numbers of both plant and animal populations in the fish tank are kept balanced.

....

....

.... **(3 marks)**

The water cycle

1 (a) Give **three** natural sources of water vapour in the atmosphere.

> The question says 'water vapour' so make sure that your answer talks about the formation of water vapour and not about other aspects of the water cycle.

....

....

....

....

.... **(3 marks)**

(b) Describe what happens when water vapour in the atmosphere condenses.

....

....

....

....

.... **(3 marks)**

2 In parts of California there is a lack of rainfall. Water has been taken from rivers and used to water lawns and golf courses. Some of these areas are suffering from drought and there are now restrictions on the number of days a week golf courses can be watered. Explain why these restrictions have been introduced.

Guided

A lot of water evaporates from golf courses so this will lead to...............................

....

....

.... **(3 marks)**

3 Sea water contains too much salt to make it potable (safe to drink). Potable water can be produced from sea water by desalination:

• sea water is evaporated by heating

• water vapour is cooled and condensed.

Give **one** advantage of desalination to people in areas where there is a drought, and **one** disadvantage.

....

....

....

....

.... **(2 marks)**

The nitrogen cycle

The diagram shows how the element nitrogen moves between living organisms and the environment.

1 Bacteria are involved in different stages of the nitrogen cycle. Which is the correct combination of processes involving bacteria?

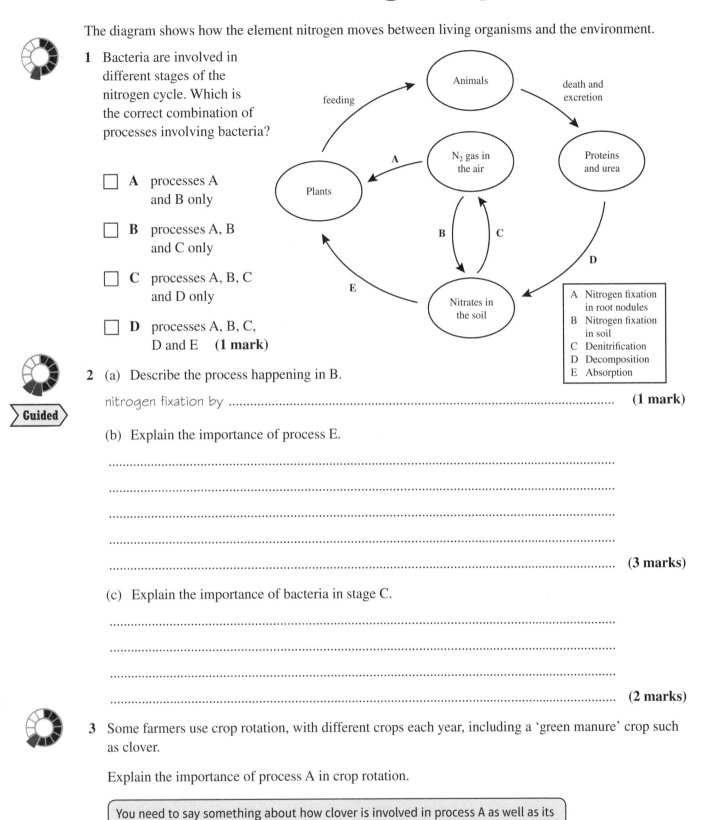

☐ A　processes A and B only

☐ B　processes A, B and C only

☐ C　processes A, B, C and D only

☐ D　processes A, B, C, D and E　(1 mark)

2 (a) Describe the process happening in B.

nitrogen fixation by .... (1 mark)

(b) Explain the importance of process E.

....

....

....

....

.... (3 marks)

(c) Explain the importance of bacteria in stage C.

....

....

....

.... (2 marks)

3 Some farmers use crop rotation, with different crops each year, including a 'green manure' crop such as clover.

Explain the importance of process A in crop rotation.

> You need to say something about how clover is involved in process A as well as its importance for crop plants.

....

....

....

....

.... (3 marks)

Extended response – Ecosystems and material cycles

Explain how fish farming and other human activities have an impact on biodiversity.

> You will be more successful in extended response questions if you plan your answer before you start writing.
>
> Try to include a number of different examples of how human activity has an impact on biodiversity. Remember that not all human activity is bad for biodiversity; try to think of some examples where human activity can increase biodiversity.

....

....

....

....

....

....

....

....

....

....

....

....

....

....

....

....

....

....

....

....

....

....

.... **(6 marks)**

Formulae

1   Which of the following is the formula for calcium carbonate?

     ☐   **A** $CaCO$

     ☐   **B** $CaCO_2$

     ☐   **C** $CaCO_3$

     ☐   **D** $CaCO_4$ **(1 mark)**

> Put a cross in **one** box. Always answer multiple-choice questions, even if you don't actually know the answer.

2   Chlorine is used to kill harmful microorganisms in drinking water. Its formula is Cl_2. Place a tick (✓) in each correct box to describe what this formula tells you.

Cl_2 tells you that:	Tick (✓)
chlorine is an element	
chlorine is a compound	
chlorine is a mixture of atoms	
chlorine exists as molecules	

           (2 marks)

3   Complete the table to show the formulae of some common substances.

Substance	water	carbon dioxide	methane	sulfuric acid	sodium
Formula					

           (5 marks)

4   State what is meant by the term '**element**'.

> **Guided**

    An element is a substance made from ....

    with the same number of ....   **(2 marks)**

5   The formula for aluminium hydroxide is $Al(OH)_3$.

    (a)   Deduce the number of elements in the formula $Al(OH)_3$.

    ....   **(1 mark)**

    (b)   Deduce the total number of atoms in the formula $Al(OH)_3$.

    ....   **(1 mark)**

6   The formula for a carbonate ion is CO_3^{2-}.

    (a)   State how you can tell that this is the formula for an ion.

    ....   **(1 mark)**

    (b)   Describe what the numbers in the formula tell you about a carbonate ion.

    ....

    ....   **(2 marks)**

Equations

1 Which of these statements describes a chemical reaction?

   ☐ **A** Reactants form from products.

   ☐ **B** Products form from reactants.

   ☐ **C** An element changes into another element.

   ☐ **D** The total mass of substances goes down.

   > Answer C cannot be correct because one element cannot change to another element in chemical reactions.

   (1 mark)

2 The word equation for the thermal decomposition of copper carbonate is:

   copper carbonate → copper oxide + carbon dioxide

   Complete the table by placing a tick (✓) in one box in each row to show if a substance is a product or a reactant in this reaction.

Substance	Reactant	Product
copper oxide		
copper carbonate		
carbon dioxide		

   (2 marks)

3 Sodium hydroxide solution reacts with dilute hydrochloric acid to form sodium chloride and water.

   Write the word equation for this reaction.

   .... **(1 mark)**

4 A teacher adds a piece of sodium metal to some water. The reaction produces sodium hydroxide solution and bubbles of hydrogen. Complete the balanced equation below to show the correct state symbols.

   > You should be able to use the state symbols (s), (l), (g) and (aq).

   $2Na(........) + 2H_2O(........) \rightarrow 2NaOH(........) + H_2(........)$ **(1 mark)**

5 Balance the following equations by adding balancing numbers in the space provided.

> Guided

   > Do not add state symbols unless you are asked for them.

   (a) $2Cu + O_2 \rightarrow .....CuO$ **(1 mark)**

   (b) $.....Al + Fe_2O_3 \rightarrow Al_2O_3 + ...Fe$ **(1 mark)**

   (c) $Mg + ...HNO_3 \rightarrow Mg(NO_3)_2 + H_2$ **(1 mark)**

   (d) $Na_2CO_3 + ...HCl \rightarrow .....NaCl + H_2O + CO_2$ **(1 mark)**

   (e) $Cl_2 + ...NaBr \rightarrow .....NaCl + Br_2$ **(1 mark)**

   (f) $.....Fe + ...O_2 \rightarrow .....Fe_2O_3$ **(1 mark)**

Hazards, risk and precautions

1 Complete the diagram below using a straight line to connect each hazard symbol to its correct description.

Guided

Symbol	Description
	flammable may easily catch fire
	oxidising agent may cause other substances to catch fire, or make a fire worse
	corrosive causes severe damage to skin and eyes
	harmful or irritant health hazard
	toxic may cause death by inhalation, ingestion or skin contact

(4 marks)

2 Hazard symbols are found on containers. Give **two** reasons why these hazard symbols are used.

1: ....

2: .... **(2 marks)**

3 Describe what is meant by the term '**hazard**'.

Guided

A hazard is something that could cause ....

or cause .... **(2 marks)**

4 Describe what is meant by the term '**risk**'.

Risk is the chance that ....

> Risk and hazard are **not** the same thing.

Guided

....

.... **(2 marks)**

5 Copper reacts with concentrated nitric acid. The reaction forms copper nitrate, water and nitrogen dioxide. Nitrogen dioxide is a toxic brown gas with an irritating odour.

(a) Give **one** suitable precaution, other than eye protection, needed for safe working in this experiment.

.... **(1 mark)**

(b) Give a reason that explains your answer to (a).

.... **(1 mark)**

Atomic structure

1 How much smaller is the nucleus of an atom compared with the overall size of the atom?

☐ **A** about 10 times smaller

☐ **B** about 100 times smaller

☐ **C** about 1000 times smaller

☐ **D** about 100 000 times smaller **(1 mark)**

2 Which of these statements correctly describes an atom?

☐ **A** Most of the mass is concentrated in the nucleus.

☐ **B** Most of the charge is concentrated in the nucleus.

☐ **C** The number of neutrons always equals the number of protons.

☐ **D** The number of electrons always equals the number of neutrons. **(1 mark)**

3 Atoms contain protons, neutrons and electrons. Place a tick (✓) in each correct box to show where these particles are found in atoms.

	Protons	Neutrons	Electrons
Nucleus			
Shells			

(2 marks)

4 Complete the table to show the relative mass and relative charge of each particle in an atom.

> **Guided**

Particle	Proton	Neutron	Electron
Relative mass		1	
Relative charge			–1

(2 marks)

5 Atoms contain equal numbers of protons and electrons. For example, a hydrogen atom contains one proton and one electron.

> Think about the charges carried by protons and electrons.

Explain why the overall charge of an atom is zero.

....

.... **(2 marks)**

6 John Dalton described his model of the atom in 1803.

Suggest a reason that explains why his model did not include protons, neutrons and electrons.

.... **(1 mark)**

Isotopes

1   State what is meant by the **mass number** of an atom.

> Guided

The mass number of an atom is the total number of ....

.... **(1 mark)**

2   An atom of an element X has an atomic number 9 and a mass number 19. How many electrons does an atom of element X contain?

☐ **A** 9

☐ **B** 10

☐ **C** 19

☐ **D** 28

(1 mark)

3   Describe, in terms of the particles in its atoms, what an element is.

> Guided

An element consists of atoms that have the same number of ....

in the nucleus, and this is different for different .... **(2 marks)**

4   Three isotopes of hydrogen are ^1_1H (hydrogen-1), ^2_1H (hydrogen-2) and ^3_1H (hydrogen-3).

> Guided

(a) Complete the table to show the numbers of protons, neutrons and electrons in an atom of each isotope.

Isotope	Protons	Neutrons	Electrons
hydrogen-1	1		1
hydrogen-2		1	
hydrogen-3			

(3 marks)

(b) Explain, in terms of the particles in the atom, why these are isotopes of the same element.

Isotopes of an element have atoms with the same number of ....

but different numbers of .... **(2 marks)**

5   Chlorine has a relative atomic mass of 35.5 but some elements have relative atomic masses that are whole numbers. Explain why the relative atomic masses of some elements are *not* whole numbers.

> Think about whether all the atoms of an element are the same.

....

....

.... **(2 marks)**

Had a go ☐   Nearly there ☐   Nailed it! ☐

Mendeleev's table

1 (a) How did Mendeleev **first** arrange the elements in his periodic table?

☐ **A** in the order of increasing number of protons in the nucleus

☐ **B** in the order of increasing reactivity with other elements

☐ **C** in the order of increasing number of isotopes

☐ **D** in the order of increasing relative atomic mass   **(1 mark)**

(b) State **one** factor, other than the one in your answer to (a), that Mendeleev used when he arranged the elements.

> What are the similarities and differences between elements?

.... **(1 mark)**

2 The diagram shows part of Mendeleev's 1871 table.

	Group						
	1	2	3	4	5	6	7
	H						
	Li	Be	B	C	N	O	F
	Na	Mg	Al	Si	P	S	Cl
	K Cu	Ca Zn	* *	Ti *	V As	Cr Se	Mn Br
	Rb Ag	Sr Cd	Y In	Zr Sn	Nb Sb	Mo Te	* I

(a) Give **one** similarity between this table and the modern periodic table.

> Remember that you will be given a periodic table in the exam. There is also one at the back of this book.

.... **(1 mark)**

(b) The * symbols in the diagram above represent gaps that Mendeleev left in his table.

(i) Give **two** other differences between this table and the modern periodic table.

1: ....

2: .... **(2 marks)**

(ii) Describe **one** useful thing that Mendeleev was able to do using information about the elements next to the gaps in his table.

....

.... **(1 mark)**

> Guided

3 Mendeleev had difficulty placing some elements. For example, the order of tellurium $^{128}_{53}$Te and iodine $^{127}_{54}$I appeared to be reversed in his table. Explain, in terms of atomic structure, why the positions of these two elements were actually correct.

Tellurium has a .... relative atomic mass than iodine does.

However, iodine atoms have .... protons than tellurium atoms do.   **(2 marks)**

The periodic table

1 How are the elements arranged in the modern periodic table?

☐ **A** in order of increasing mass number

☐ **B** in order of increasing atomic number

☐ **C** in order of increasing nucleon number

☐ **D** in order of increasing numbers of electron shells **(1 mark)**

2 The positions of five elements (**A**, **B**, **C**, **D** and **E**) are shown in the periodic table below. These letters are **not** the chemical symbols for these elements.

```
   1  2                          3  4  5  6  7  0
  ┌──┬──┐        ┌──┐        ┌──┬──┬──┬──┬──┬──┐
  │A │  │        │  │        │  │  │  │  │  │  │
  ├──┼──┴──┬──┬──┬──┬──┬──┬──┬──┼──┼──┼──┼──┼──┤
  │B │     │  │  │  │  │  │  │  │  │  │  │  │E │
  ├──┼─────┼──┼C─┼──┼──┼──┼──┼──┼──┼──┼──┼──┼──┤
  │  │     │  │C │  │  │  │  │  │  │  │  │  │  │
  ├──┼─────┼──┼──┼──┼──┼──┼──┼──┼D─┼──┼──┼──┼──┤
  │  │     │  │  │  │  │  │  │  │D │  │  │  │  │
  └──┴─────┴──┴──┴──┴──┴──┴──┴──┴──┴──┴──┴──┴──┘
```

(a) State the name given to a vertical column in the periodic table.

.... **(1 mark)**

(b) Give the letters of **two** elements that have similar chemical properties to each other.

.... **(1 mark)**

(c) Give the letters of **all** the metal elements.

> There are more metallic elements in the periodic table than non-metallic elements.

.... **(1 mark)**

(d) Give the letters of **two** elements in the same period.

.... **(1 mark)**

3 The meaning of the term '**atomic number**' has changed over time because of the discovery of subatomic particles.

Guided

(a) Give the meaning of the term '**atomic number**' as Mendeleev might have understood it in the nineteenth century.

the position of ....

.... **(1 mark)**

(b) Give the modern meaning of the term '**atomic number**'.

the number of ....

in an atom's .... **(2 marks)**

Electronic configurations

1   The diagram shows a lithium atom. It is not drawn to scale.

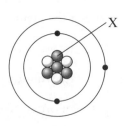

(a) State the electronic configuration of lithium.

> Count the number of electrons in each shell in the diagram.

.... **(1 mark)**

Guided

(b) Deduce the name of the shaded particle labelled **X**, and explain your answer.

There are three electrons, so there must be three ....

so the four shaded circles must be .... **(2 marks)**

(c) Oxygen (atomic number 8) has eight electrons in its atoms.

> The first electron shell can hold only a maximum of two electrons.

Draw a diagram to show the arrangement of electrons in an oxygen atom.

> You need to show both electron shells and all eight electrons, but you can show the nucleus as a single dot.

(2 marks)

Guided

2   The table shows some information about two non-metal elements, fluorine and chlorine.

Non-metal	Atomic number	Electronic configuration
F	9	2.7
Cl	17	2.8.7

(a) Explain, in terms of electronic configurations, why fluorine and chlorine are placed in Group 7.

Both have ....

in their .... **(2 marks)**

(b) Explain, in terms of electronic configurations, why fluorine and chlorine are **not** in the same period.

Fluorine has ....

but chlorine has .... **(2 marks)**

Guided

3   Deduce the electronic configurations of the following elements.

(a) phosphorus (atomic number 15)

2.8 .... **(1 mark)**

(b) calcium (atomic number 20)

.... **(1 mark)**

4   State and explain the number of the Group in which helium (electronic configuration 2) is placed.

....

.... **(2 marks)**

Ions

1 Which of the following statements correctly describes the formation of an ion?

You can quickly narrow the alternatives if you know the correct name for each type of ion, or how it forms.

☐ **A** Positively charged ions, called cations, form when atoms or groups of atoms gain electrons.

☐ **B** Positively charged ions, called anions, form when atoms or groups of atoms lose electrons.

☐ **C** Negatively charged ions, called cations, form when atoms or groups of atoms lose electrons.

☐ **D** Negatively charged ions, called anions, form when atoms or groups of atoms gain electrons.

(1 mark)

> Guided

2 The atomic number of magnesium, Mg, is 12. The symbol for a magnesium ion is Mg^{2+}.

(a) Deduce the number of electrons in a magnesium **ion**.

12 − ...... = ... **(1 mark)**

(b) The electronic configuration for a calcium atom is 2.8.8.2. Write the electronic configuration of a calcium **ion**, Ca^{2+}.

.... **(1 mark)**

> Guided

3 Complete the table to show the numbers of protons, neutrons and electrons in each ion.

Maths skills Work out the number of electrons in an atom, then add or subtract electrons according to the charge shown.

Ion	Atomic number	Mass number	Protons	Neutrons	Electrons
N^{3-}	7	15	7	8	10
K^+	19	39			
Ca^{2+}	20	40			
S^{2-}	16	32			
Br^-	35	80			

(4 marks)

4 The diagram shows the formation of a sodium ion, Na^+, from a sodium atom.

Draw a similar diagram to show the formation of a chloride ion, Cl^-, from a chlorine atom.

Your diagram should look similar to the one above. However, the electronic configuration of a chlorine atom is 2.8.7 and a chloride ion forms when a chlorine atom gains one electron.

(3 marks)

Formulae of ionic compounds

1 The formula of a sodium ion is Na^+. The formula of a phosphate ion is PO_4^{3-}. Which of the following is the formula for sodium phosphate?

> Answer A cannot be correct because the sodium ion has fewer charges than the phosphate ion.

☐ A $NaPO_4$ ☐ B $Na(PO_4)_3$ ☐ C Na_2PO_4 ☐ D Na_3PO_4 **(1 mark)**

2 Complete the table to show the formulae of the compounds produced by each pair of ions.

> You need to know the formulae of common ions. This helps you work out the formulae of ionic substances.

> **Maths skills** An ionic compound contains equal numbers of positive and negative *charges*, but not always equal numbers of positive and negative *ions*. Look at the completed examples to help you.

	Cl^-	S^{2-}	OH^-	NO_3^-	SO_4^{2-}
K^+				KNO_3	
Ca^{2+}			$Ca(OH)_2$		$CaSO_4$
Fe^{3+}		Fe_2S_3			
NH_4^+	NH_4Cl				

(15 marks)

3 Magnesium ribbon burns in air. It reacts with oxygen to produce magnesium oxide.

(a) Balance this equation for the reaction.

......$Mg + O_2 \rightarrow$ ......MgO **(1 mark)**

(b) Magnesium nitride is also formed, as some of the hot magnesium reacts with nitrogen in the air.

> Think about how many electrons a nitrogen atom must lose or gain to obtain a full outer shell.

(i) Nitrogen is in Group 5. Suggest reasons that explain why the formula for a nitride ion is N^{3-}.

....

.... **(2 marks)**

(ii) Write the formula for magnesium nitride.

> The formula for a magnesium ion is Mg^{2+}.

.... **(1 mark)**

(iii) Explain why the NO_3^- ion is called the nitrate ion, but the N^{3-} ion is called the nitride ion.

....

.... **(2 marks)**

4 Complete the table to show the names of the ions.

> Remember to use the endings -ide and -ate correctly. Look again at question 3 (b) (iii) to help you.

	S^{2-}	SO_4^{2-}	Cl^-	ClO_3^-
Name				

(4 marks)

> S is the chemical symbol for sulfur and Cl is the chemical symbol for chlorine.

Properties of ionic compounds

1 Which statement about the formation of ionic compounds, such as sodium chloride, is correct?

☐ **A** Electrons are transferred from metal atoms to non-metal atoms, producing cations and anions.

☐ **B** Electrons are transferred from cations to anions, producing metal atoms and non-metal atoms.

☐ **C** Electrons are shared between metal atoms and non-metal atoms.

☐ **D** Electrons are shared between cations and anions. **(1 mark)**

2 Ionic compounds have a lattice structure.

(a) Complete the diagram, using the symbols + and –, to show the positions of positive and negative ions in an ionic lattice.

> Remember that opposite charges will attract each other and like charges will repel.

> **Maths skills** You should be able to visualise and represent 2D and 3D forms, including 2D models of 3D objects. **(1 mark)**

Guided (b) Describe what ionic bonds are.

There are strong ....

between .... **(2 marks)**

3 Explain why ionic compounds have high melting points and boiling points.

> Bonds between the particles in an ionic substance must be broken during melting and boiling. Think about whether this involves a relatively low or high amount of energy, and why.

.... **(2 marks)**

4 Ionic compounds such as sodium chloride can conduct electricity in some situations.

(a) Complete the table by placing a tick (✓) in each **correct** box to show where ionic compounds conduct electricity.

> You do not need to tick all the boxes.

Ionic compound is:	solid	liquid	dissolved in water
conducts electricity			

(1 mark)

(b) State what the ions in an ionic compound must be able to do for it to conduct electricity.

.... **(1 mark)**

Had a go ☐   Nearly there ☐   Nailed it! ☐

Covalent bonds

1 What are the typical sizes of atoms and small molecules?

> **Maths skills** The quantities are shown in standard form. For example, 10^{-3} is greater than 10^{-6}.

	Atoms	Molecules
☐ A	10^{-10} m	10^{-11} m
☐ B	10^{-9} m	10^{-12} m
☐ C	10^{-10} m	10^{-9} m
☐ D	10^{-12} m	10^{-9} m

> Answer A cannot be correct because it shows atoms as being larger than small molecules.

(1 mark)

2 Explain how a covalent bond forms.

> How many electrons are involved in a covalent bond?

Guided

A covalent bond forms when ....

between .... **(2 marks)**

3 Hydrogen reacts with fluorine to form hydrogen fluoride: $H_2 + F_2 \rightarrow 2HF$

The electronic configuration of hydrogen is 1 and the electronic configuration of fluorine is 2.7.

(a) Describe what the structure, H–F, tells you about a hydrogen fluoride molecule.

....

.... **(2 marks)**

(b) A dot-and-cross diagram for a molecule of fluorine, F_2, is shown below.

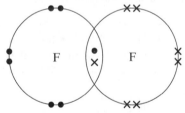

> Show each chemical symbol. Show one atom's electrons as dots and the other atom's electrons as crosses.

Draw a dot-and-cross diagram for a molecule of hydrogen fluoride, HF. Show the outer electrons only.

(2 marks)

4 Oxygen atoms have six electrons in their outer shell.
Draw a dot-and-cross diagram for an oxygen molecule, O_2.
Show the outer electrons only.

> The displayed formula for oxygen is O=O.

(2 marks)

Simple molecular substances

1 Carbon dioxide, CO_2, is found in the air. Why does it have a low boiling point?

> Answer A cannot be correct because covalent bonds are strong.

☐ **A** There are weak covalent bonds between carbon atoms and oxygen atoms.

☐ **B** There are weak forces of attraction between carbon atoms and oxygen atoms.

☐ **C** There are weak forces of attraction between carbon dioxide molecules.

☐ **D** There are weak covalent bonds between carbon dioxide molecules. **(1 mark)**

2 The table shows the properties of four different substances (**A**, **B**, **C** and **D**).

Substance	Melting point (°C)	Conducts electricity when solid?	Conducts electricity when liquid?	Solubility in water (g per 100 g of water)
A	290	no	yes	43
B	−39	yes	yes	0
C	−95	no	no	0.001
D	660	yes	yes	0

(a) State which substance (**A**, **B**, **C** or **D**) is a simple molecular substance.

.... **(1 mark)**

(b) Explain your answer to (a).

> Which of the properties of your chosen substance are typical of simple molecular substances?

....

.... **(2 marks)**

3 Sulfur hexafluoride, SF_6, exists as simple molecules. It is used as an insulating gas for electrical equipment.

(a) Explain why sulfur hexafluoride does not conduct electricity.

> Substances that conduct electricity have electrically charged particles that are free to move around. Think about whether simple molecules are electrically charged or contain electrons that are free to move.

....

.... **(2 marks)**

> **Guided**

(b) Suggest reasons that explain why sulfur hexafluoride does not dissolve in water.

The intermolecular forces between ....

....

are weaker than those between ....

and those between .... **(3 marks)**

Giant molecular substances

1 Silica, SiO_2, does not dissolve in water. It does not conduct electricity, even when molten, and its melting point is very high.

Which statement describes a molecule of silica?

> Answer D cannot be correct because metals, not molecules, contain metallic bonds.

☐ **A** a giant molecule with ionic bonds

☐ **B** a giant molecule with covalent bonds

☐ **C** a simple molecule with covalent bonds

☐ **D** a simple molecule with metallic bonds

(1 mark)

2 The diagrams show the structures of diamond and graphite.

> **Maths skills** You should be able to visualise and represent 2D and 3D forms, including 2D representations of 3D objects.

diamond   graphite

(a) Name the element with atoms that form both diamond and graphite.

.... **(1 mark)**

(b) State the maximum number of bonds present between each atom in a molecule of diamond.

> Count the bonds between the atoms in the diagram of diamond. What is the highest number you get?

.... **(1 mark)**

(c) Name the type of structure shown in both diagrams.

.... **(1 mark)**

3 Refer to structure and bonding in your answers to the following questions.

> **Guided**

(a) Explain why graphite is suitable for use as a lubricant. 〔Lubricants must be slippery.〕

The layers in graphite can ....

because .... **(2 marks)**

(b) Explain why graphite is used to make electrodes.

> You need to explain why graphite can conduct electricity, just as metals can.

Atoms in graphite can form only three ....

so graphite has .... **(2 marks)**

(c) Explain why diamond is suitable for use in cutting tools.

> You need to explain why diamond is very hard.

Diamond has a .... structure, and its atoms

are joined by many .... **(2 marks)**

Other large molecules

1   Ethene, C_2H_4, can be made into a polymer. What is the name of this polymer?

☐ **A** poly(ethanol)   ☐ **C** poly(ethene)

☐ **B** poly(ethane)   ☐ **D** poly(ethyne)   **(1 mark)**

2   The diagram is a model of a section of a simple polymer.

(a) Name the element with atoms represented by the larger, dark-grey balls in the diagram.

.... **(1 mark)**

(b) Name the type of bonding present in a molecule of this polymer.

.... **(1 mark)**

3   Fullerenes are forms of carbon that include hollow balls, such as buckminsterfullerene, C_{60}.

(a) Explain why buckminsterfullerene is a simple molecule, rather than a giant covalent substance.

....

....

.... **(2 marks)**

Guided

(b) Explain, in terms of its structure and bonding, why buckminsterfullerene has a much lower melting point than graphite.

> The strong covalent bonds between the carbon atoms in these molecules do *not* break during melting.

Buckminsterfullerene has a .... structure

so it has weak .... that are easily overcome.   **(2 marks)**

4   Graphene is a form of carbon. It is a good conductor of electricity and has a very high melting point.

The diagram is a model of part of the structure of graphene.

Explain, in terms of its structure and bonding, why graphene has a very high melting point.

> Include the type of bonds that must be broken during melting.

....

....

.... **(3 marks)**

Metals

1 Which of the following correctly describes two typical properties of metals?

☐ **A** shiny with high densities   ☐ **C** dull with low densities

☐ **B** shiny with low densities   ☐ **D** dull with high densities   **(1 mark)**

2 Metal elements and non-metal elements have different typical properties.

Complete the table below by placing a tick (✓) in each correct box.

	Low melting points	High melting points	Good conductors of electricity	Poor conductors of electricity
Metals				
Non-metals				

(4 marks)

3 Metals are insoluble in water. Some metals react with water, forming soluble hydroxides and hydrogen. For example, a piece of sodium reacts with water to produce sodium hydroxide and hydrogen.

(a) State why fizzing is observed during this reaction.

.... **(1 mark)**

(b) Suggest reasons that explain why the piece of sodium seems to dissolve in water.

....

.... **(2 marks)**

4 Copper is a metal used in electricity cables. It is a good conductor of electricity and is malleable (it will bend without shattering). The diagram is a model for the structure of copper. Each circle is a copper ion.

(a) State **two** improvements to the diagram that will make it a more accurate model of the structure of copper.

> Remember that metal atoms form positively charged ions by losing electrons.

(i) ....

(ii) .... **(2 marks)**

Guided (b) Explain why copper is malleable.

It has layers of ....

which can .... **(2 marks)**

(c) Explain why copper is a good conductor of electricity.

> A substance conducts electricity if it contains charged particles that are free to move around.

....

.... **(2 marks)**

Limitations of models

1 The formula of a substance can be given in different ways.

Which row (**A**, **B**, **C** or **D**) correctly shows the different formulae for ethene?

> Answer A cannot be correct because it describes ethane, not ethene.

	Molecular formula	Empirical formula	Structural formula
☐ **A**	C_2H_6	CH_3	CH_3CH_3
☐ **B**	C_2H_4	CH_2	$CH_2=CH_2$
☐ **C**	CH_2	C_2H_4	$CH_2=CH_2$
☐ **D**	$CH_2=CH_2$	C_2H_4	CH_2

(1 mark)

2 The diagrams (**A**, **B**, **C** and **D**) show four different models for a molecule of methane, CH_4.

A	B	C	D
H—C—H with H above and H below (structure)	dot-and-cross diagram with central C and four H	ball-and-stick model	space-filling model
Structure	Dot-and-cross diagram	Ball-and-stick model	Space-filling model

State the letters (**A**, **B**, **C** or **D**) for all the models that:

> You may need to identify more than one model in your answers.

(a) show the covalent bonds present in a molecule .... **(1 mark)**

(b) identify the elements present in a molecule .... **(1 mark)**

(c) represent the three-dimensional shape of a molecule .... **(1 mark)**

(d) show the electrons involved in bonding .... **(1 mark)**

(e) show the relative sizes of each atom in a molecule .... **(1 mark)**

Guided

3 A student wants to represent a water molecule. She decides to draw a dot-and-cross diagram rather than a ball-and-stick model because she finds this easier to do.

> Think about what a ball-and-stick model shows that a dot-and-cross diagram does not.

(a) Give **one** strength of a ball-and-stick model compared with a dot-and-cross diagram.

Unlike a dot-and-cross diagram, a ball-and-stick model ....

.... **(1 mark)**

(b) Other than the student's reason, give two weaknesses of a ball-and-stick model compared with a dot-and-cross diagram.

> Think about what a dot-and-cross model shows that a ball-and-stick model does not.

Unlike a dot-and-cross diagram, a ball-and-stick model does not show ....

....

or .... **(2 marks)**

99

Relative formula mass

Use the relative atomic masses, A_r, in the table below when you answer the questions.

Element	Al	Ca	Cl	Cu	H	N	O	S
A_r	27	40	35.5	63.5	1	14	16	32

> If relative atomic masses are not given in the question, you can find them in the periodic table.

> **Guided**

1 Calculate the relative formula mass, M_r, of each of the following substances.

> You do not need to show your working out, but it will help you to check the accuracy of your answers.

(a) chlorine, Cl_2

> You do not need to show a decimal point in your answer to this question.

$2 \times 35.5 =$ .... **(1 mark)**

(b) water, H_2O

$(2 \times 1) + 16 = 2 + 16 =$ .... **(1 mark)**

(c) sulfur dioxide, SO_2.. **(1 mark)**

(d) aluminium oxide, Al_2O_3 .... **(1 mark)**

(e) ammonium chloride, NH_4Cl.. **(1 mark)**

> Do not round the answer to this question to a whole number.

(f) calcium chloride, $CaCl_2$ .... **(1 mark)**

(g) aluminium chloride, $AlCl_3$ .... **(1 mark)**

> **Guided**

2 Calculate the relative formula mass, M_r, of each of the following substances.

(a) calcium hydroxide, $Ca(OH)_2$

$16 + 1 = 17, 17 \times 2 = 34, 40 + 34 =$ .... **(1 mark)**

> 🖩 **Maths skills** You could also enter the calculation into your calculator as: $40 + [2 \times (16 + 1)] =$

(b) aluminium hydroxide, $Al(OH)_3$ .... **(1 mark)**

(c) calcium nitrate, $Ca(NO_3)_2$ .... **(1 mark)**

(d) ammonium sulfate, $(NH_4)_2SO_4$ .... **(1 mark)**

(e) aluminium sulfate, $Al_2(SO_4)_3$ .... **(1 mark)**

Practical skills Empirical formulae

1   A student carries out an experiment to determine the empirical formula of magnesium oxide. He heats a piece of magnesium ribbon in a crucible. He continues until the contents of the crucible stop glowing.

The table shows his results.

Object	Mass (g)
empty crucible and lid	20.25
crucible, lid and contents before heating	20.49
crucible, lid and contents after heating	20.65

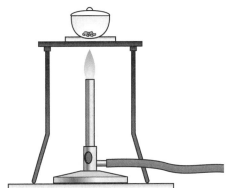

(a) Suggest a reason that explains why:

(i)   the student continued heating until the glowing stopped.

.... (1 mark)

(ii) the student briefly lifted the lid a few times during the experiment.

.... (1 mark)

(b) The hot crucible is a hazard. Explain one precaution to control the risk of harm from this hazard.

> Say what the student should do to avoid being harmed, and what harm this will prevent.

....

.... (2 marks)

2   Calculate the empirical formula of magnesium oxide using the student's results.

> Guided

(A_r of Mg = 24 and A_r of O = 16)

mass of magnesium used = 20.49 g − 20.25 g = 0.24 g

mass of oxygen reacted = 20.65 g − 20.49 g = ...

$$\text{Mg} \qquad\qquad \text{O}$$

$\dfrac{0.24}{24} = 0.010$ $\dfrac{.........}{16} = ...$

> Divide the mass of each element by its A_r.

$\dfrac{0.010}{........} = ...$ $\dfrac{.........}{........} = ...$

> Divide both numbers by the smallest number to find the ratio.

Empirical formula is ....

> Write down the empirical formula.   (4 marks)

3   The empirical formula of a sample of gas is NO_2. Its relative formula mass, M_r, is 92.

> Guided

Deduce the molecular formula of the gas. (A_r of N = 14 and A_r of O = 16)

M_r of NO_2 = 14 + (2 × 16) = ...

factor needed = $\dfrac{92}{........}$ =

> Maths skills Calculate the M_r of NO_2. Then work out how many times this will go into 92. Multiply each number in the empirical formula by this factor to obtain the molecular formula.

Molecular formula is ....

(2 marks)

Conservation of mass

1   Sodium chloride solution reacts with silver
    nitrate solution. Sodium nitrate solution and
    a white precipitate of solid silver chloride form:
    $NaCl(aq) + AgNO_3(aq) \rightarrow NaNO_3(aq) + AgCl(s)$

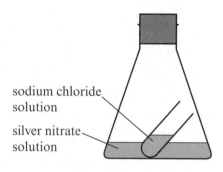

sodium chloride
solution

silver nitrate
solution

A student investigates the change in mass
during this reaction. He sets up the apparatus
shown in the diagram, then shakes the flask
to mix the two solutions.

(a) State whether the reaction happens in a closed or a non-enclosed system.
     Give a reason for your answer.

type of system: ....

reason: .... **(1 mark)**

(b) The student records the mass of the flask and its contents before and after the reaction.

   (i)   What happens to the mass during the reaction? <u>Underline</u> the correct answer.

                    It increases. | It decreases. | It stays the same.   **(1 mark)**

   (ii) Give a reason for your answer to (i).

         .... **(1 mark)**

2   Copper carbonate decomposes, when heated, to form copper oxide and carbon dioxide:

    $CuCO_3(s) \rightarrow CuO(s) + CO_2(g)$

    | You do not need to calculate relative
    | formula masses for this question.

    8.2 g of copper carbonate formed 5.3 g of copper oxide.
    Calculate the mass of carbon dioxide produced. .... **(1 mark)**

3   Sodium reacts with chlorine to form sodium chloride: $2Na(s) + Cl_2(g) \rightarrow 2NaCl(s)$

> Guided

    Calculate the maximum mass of sodium chloride that can be made from 142 g of chlorine.

    (M_r of Cl_2 = 71 and M_r of NaCl = 58.5)

    $(1 \times 71) = 71\,g$ of Cl_2 makes $(2 \times 58.5) = 117\,g$ of NaCl

    $142\,g$ of Cl_2 makes $117 \times (142/71)\,g$ of NaCl

    = ... g   **(1 mark)**

4   Magnesium reacts with oxygen to form magnesium oxide: $2Mg(s) + O_2(g) \rightarrow 2MgO(s)$

> Guided

    Calculate the maximum mass of magnesium
    oxide that can be made from 12.6 g of oxygen.

    | Remember to calculate the relative
    | formula mass, M_r, of oxygen gas
    | and magnesium oxide first.

    (A_r of O = 16 and A_r of Mg = 24)

    M_r of O_2 = ... M_r of MgO = ...

    .... g of O_2 makes .... g of MgO

    12.6 g of O_2 makes .... g of MgO

    = ... g of MgO   **(3 marks)**

Concentration of solution

1   Sea water contains dissolved salts.

Which row (**A**, **B**, **C** or **D**)
correctly describes the
components of sea water?

	Solute	Solvent	Solution
☐ **A**	water	salt	sea water
☐ **B**	salt	sea water	water
☐ **C**	sea water	salt	water
☐ **D**	salt	water	sea water

(1 mark)

> **Guided**

2   Calculate the following volumes in dm^3.

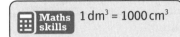
Maths skills $1\,dm^3 = 1000\,cm^3$

(a) $2500\,cm^3$

$$volume = \frac{2500}{1000} = \ldots\ldots\ldots\ldots\ dm^3$$

(1 mark)

(b) $500\,cm^3$

.... **(1 mark)**

(c) $25\,cm^3$

.... **(1 mark)**

3   Calculate the concentrations of the following solutions in $g\,dm^{-3}$:

(a) $50\,g$ of sodium hydroxide dissolved in $2\,dm^3$ of water

.... **(1 mark)**

(b) $14.6\,g$ of hydrogen chloride dissolved in $0.400\,dm^3$ of water

.... **(1 mark)**

(c) $0.25\,g$ of glucose dissolved in $0.050\,dm^3$ of water.

.... **(1 mark)**

> **Guided**

4   A student dissolves $10\,g$ of copper sulfate in $250\,cm^3$ of water. Calculate the
    concentration of the solution formed in $g\,dm^{-3}$.

$$concentration = \left(\frac{10}{250}\right) \times 1000 = \ldots\ldots\ldots\ldots\ldots\ldots\ldots\ldots\ldots\ldots\ldots\ldots\ldots\ldots\ldots$$ **(1 mark)**

5   A student dissolves $2.0\,g$ of silver nitrate in $125\,cm^3$ of water. Calculate the
    concentration of the solution formed in $g\,dm^{-3}$.

.... **(1 mark)**

6   A school technician wants to make $2.5\,dm^3$ of a $40\,g\,dm^{-3}$ aqueous solution of
    sodium hydroxide.

(a) Describe the meaning of the term '**aqueous solution**'.

.... **(1 mark)**

(b) Calculate the mass of sodium hydroxide that the
     technician must dissolve to make her solution.

Maths skills   Rearrange this equation.

concentration in $g\,dm^{-3}$ =
mass of solute in volume of solution in dm^3

.... **(1 mark)**

Extended response – Types of substance

Graphite and diamond are two different forms of carbon. The table shows some information about their properties. Copper, a soft metal used in electrical cables, is included for comparison.

Substance	Relative hardness	Relative electrical conductivity
graphite	10	10^8
copper	100	10^{10}
diamond	10 000	1

Higher values mean harder or better at conducting electricity.

Graphite is used to make electrodes and as a lubricant. Diamond is used in cutting tools.

Describe, using information from the table, the properties of graphite and diamond that make them suitable for these uses. Explain these properties in terms of the bonding and structure present.

You should be able to describe the structures of graphite and diamond.

It may help if you plan your answer before you start. For example, write separate answers about graphite and diamond. Include each given use and, using information from the table, the property important to that use. Make sure that you then explain how the substance's bonding and structure give it that property.

....

....

....

....

....

....

....

....

....

....

....

....

(6 marks)

Quick, labelled diagrams showing the structures of diamond and graphite may help your explanations.

States of matter

1 Iodine crystals become a purple vapour when they are warmed. What is the name for this state change?

> The crystals are in the solid state and the vapour is in the gas state.

☐ **A** melting   ☐ **B** boiling   ☐ **C** subliming   ☐ **D** condensing   **(1 mark)**

2 Most substances can exist in the solid, liquid or gas states. Give the name of each state change below.

(a) liquid to solid

.... **(1 mark)**

(b) gas to liquid

.... **(1 mark)**

3 Water changes to steam when it is heated. State why this is a **physical** change.

.... **(1 mark)**

4 Particles are arranged in different ways in each state of matter. Place a tick (✓) in each correct box below.

State of matter	Particles are:			
	close together	far apart	randomly arranged	regularly arranged
solid				
liquid				
gas				

(3 marks)

5 (a) Name the state in which the particles move the fastest.

.... **(1 mark)**

(b) Particles in all states have some stored energy. Name the state in which the particles have the least stored energy, and justify your answer.

....

.... **(2 marks)**

6 The melting point of substance **X** is −114 °C and its boiling point is 78 °C. Predict its state at −30 °C.

> −30 °C is above −114 °C and below 78 °C.

.... **(1 mark)**

7 Describe what happens to the arrangement, and movement, of particles when a substance changes from the liquid state to the solid state.

> Guided

The arrangement changes from ....

and the movement changes from ....

.... **(2 marks)**

Pure substances and mixtures

1 Substances can be pure or they can be mixtures. Which of the following is a mixture?

   ☐ **A** carbon   ☐ **C** carbon dioxide

   ☐ **B** oxygen   ☐ **D** carbon dioxide solution   **(1 mark)**

2 The label on a carton of orange juice describes the contents as 'pure'. Explain why the orange juice is a mixture, rather than a pure substance, in the scientific sense.

   ....

   ....

   ....   **(3 marks)**

3 (a) Explain, in terms of subatomic particles, why sodium, Na, and chlorine, Cl_2, are two different elements.

> Write your answer in terms of a named subatomic particle found in the nucleus of all atoms.

   The atoms of an element all have the same ....

   but atoms of Na and Cl_2 have different ....   **(2 marks)**

   (b) Explain why sodium chloride, NaCl, is defined as a compound.

> Think about how many elements a compound contains, and whether or how these are joined together.

   ....

   ....   **(2 marks)**

4 A student investigates three samples of water. She transfers $25 \ cm^3$ of each sample to weighed evaporating basins. She then heats the basins until all the water has evaporated, lets them cool and weighs them again. The table shows the student's results.

Water sample	Mass of basin before adding water (g)	Mass of basin after evaporating water (g)	Difference in mass (g)
A	73.20	73.05	0.15
B	72.85	72.61	
C	74.43	74.40	

   (a) Complete the table to show the difference in mass for water samples **B** and **C**.   **(1 mark)**

   (b) State whether any of the water samples are pure, and justify your answer.

   ....

   ....   **(2 marks)**

5 Solders are alloys used to join copper pipes or electrical components together. Some 'lead-free' solders are mixtures of tin and silver. The table shows the melting points of tin, silver and a lead-free solder.

Substance	Melting point (°C)
tin	232
silver	962
lead-free solder	220–229

   Explain how the data show that the lead-free solder is a mixture.

   ....

> Look at how the melting points are shown for each substance.

   ....

   ....   **(2 marks)**

Distillation

1 Which of the following is a suitable method to separate a mixture of two miscible liquids?

> Miscible liquids mix completely with each other.

☐ **A** filtration

☐ **B** simple distillation

☐ **C** fractional distillation

☐ **D** paper chromatography   **(1 mark)**

2 The apparatus shown in the diagram is used to separate the components of an ink.

(a) (i) Name the apparatus labelled **X**.

> Water vapour goes in and droplets of water come out.

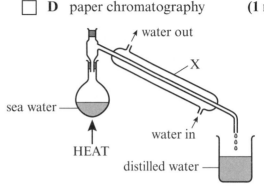

.... **(1 mark)**

(ii) State what happens to the temperature of the vapour from the sea water as it passes through apparatus **X**.

.... **(1 mark)**

 Guided

(b) Cold water passes through apparatus **X**. Explain what happens to the temperature of this water.

The temperature of the water .... because ....

....

.... **(2 marks)**

3 Ethanol boils at 78 °C and water boils at 100 °C. The apparatus shown in the diagram is used to separate a mixture of ethanol and water.

(a) Explain which liquid, ethanol or water, is collected first.

> Look again at the information given to you in the question.

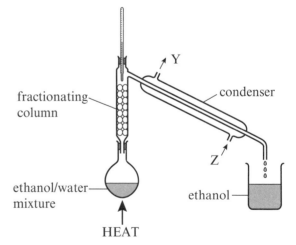

....

.... **(2 marks)**

(b) Give a reason that explains why the cold water supply should be connected at **Z** rather than **Y**.

> 🧪 **Practical skills** Hot vapour from the ethanol–water mixture will pass from left to right through this apparatus.

....

.... **(1 mark)**

Filtration and crystallisation

1   A student filters a mixture of sand, salt and water. He collects the liquid that passes through the filter paper. Complete the table by placing a tick (✓) in the box against each correct statement.

Statement	Tick (✓)
The liquid is water.	
The liquid is the filtrate.	
The salt is left behind as a residue.	
The sand is left behind as a residue.	

> Do not place a cross against the incorrect statement(s) – you are asked to place only ticks in the table.

(2 marks)

> **Guided**

2   Potassium iodide solution reacts with lead nitrate solution. A mixture of potassium nitrate solution and insoluble yellow lead iodide forms. When this is filtered, solid lead iodide remains on the filter paper.

> Use the information in the question to decide whether each substance is solid, liquid or gas, or in aqueous solution. Few balanced equations use all four state symbols.

(a)   Balance the equation below, and give the state symbols for each substance.

........$KI(aq)$ + $Pb(NO_3)_2$(........) → ....KNO_3(........) + PbI_2(........)   **(2 marks)**

(b)   The yellow lead iodide is washed, with distilled water, while it is on the filter paper.

(i)   State why the lead iodide does not pass through the filter paper.

.... **(1 mark)**

(ii)   Suggest a reason that explains why the lead iodide is washed.

.... **(1 mark)**

3   A student decides to make pure, dry copper chloride crystals. She adds an excess of insoluble copper carbonate to dilute hydrochloric acid. Copper chloride solution forms.

(a)   Name a suitable method to remove excess copper carbonate from the copper chloride solution.

> If a substance is in excess, some of it is left over when the reaction finishes.

.... **(1 mark)**

(b)   The student leaves her copper chloride solution in a dish on a windowsill. Crystals gradually form.

(i)   Describe what happens to the solution as the crystals of copper chloride form.

....

....

.... **(2 marks)**

(ii)   Give a reason that explains why the student pours away the remaining solution, and then pats the crystals with filter paper.

.... **(1 mark)**

(c)   The student could heat her copper chloride solution instead of leaving it on a windowsill. Describe one step that she should take to obtain large crystals rather than small ones.

> Think about how she needs to carry out this heating so she gets large crystals.

....

.... **(1 mark)**

Paper chromatography

1 Paper chromatography is used to determine whether an orange squash drink, **O**, contains an illegal food colouring, **X**.

Spots of each substance, and spots of three legal food colourings (**A**, **B** and **C**), are added to chromatography paper. The diagram shows the result of the chromatography experiment.

(a) Suggest a reason that explains why the start line is drawn using a pencil, rather than using ink.

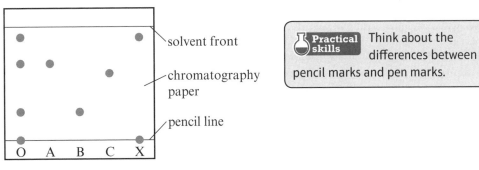

solvent front

chromatography paper

pencil line

O A B C X

> **Practical skills** Think about the differences between pencil marks and pen marks.

.... **(1 mark)**

(b) Explain whether the orange squash, **O**, is a pure substance or a mixture.

> State if the orange squash is a pure substance or a mixture, then say why.

....

.... **(2 marks)**

(c) Identify which legal food colourings (**A**, **B** or **C**) are present in the orange squash.

.... **(1 mark)**

(d) Explain whether the orange squash contains the illegal food colouring, **X**.

....

.... **(2 marks)**

(e) Suggest a reason that explains why one of the substances in **X** remains on the pencil line.

> Remember that the paper is the stationary phase and the solvent the mobile phase, which runs through it.

.... **(1 mark)**

2 The diagram shows a chromatogram of a dye.

Calculate the R_f value of the spot in the chromatogram.

$R_f = \dfrac{\text{distance travelled by a spot}}{\text{distance travelled by the solvent front}}$

solvent front

start line

(2 marks)

 Practical skills

Investigating inks

1   Paper chromatography is used to separate mixtures of coloured substances, such as those in inks.

(a) State two measurements that must be made so that an R_f value can be calculated.

> These are distances that must be measured on the finished chromatogram.

....

.... **(2 marks)**

(b) Suggest reasons that explain why these measurements are recorded to the nearest millimetre, rather than the nearest centimetre.

> Practical skills You must be able to use appropriate apparatus to make and record a range of measurements accurately.

....

.... **(2 marks)**

2   Simple distillation is used to separate a solvent from a solution.

(a) Suggest reasons that explain why the solution must be heated **gently**.

> This is not to do with a flammable solvent that might catch fire.

....

.... **(2 marks)**

(b) During distillation, the solvent vapour may condense slowly without the use of a condenser. State one hazard caused by carrying out distillation without a condenser.

> Remember that a hazard could harm something or someone, or could affect someone's health.

.... **(1 mark)**

3   A student investigates the composition of a sample of ballpoint pen ink. He uses propanone for the mobile phase in his paper chromatography. The diagram shows part of the label on a bottle of propanone.

(a) State **one** hazard of propanone, shown by the label but not described in words.

> Irritating to eyes.
> May cause skin dryness.
> Vapour causes dizziness.
> Propanone CH_3COCH_3

.... **(1 mark)**

> Guided

(b) Explain **two** precautions, other than eye protection, to control the risk of harm from propanone.

> You need to be able to use gases, liquids and solids safely and carefully.

The student should work in a fume cupboard because ....

....

The propanone causes skin dryness, so the student should ....

.... **(2 marks)**

Drinking water

1 Waste water and ground water can be treated to make it safe to drink. Which word correctly describes water that is safe for us to drink?

☐ **A** potable

☐ **C** edible

☐ **B** fresh

☐ **D** filtered

(1 mark)

2 Chlorine is a toxic gas that dissolves in water.

(a) State why chlorine is used in water treatment.

> Your answer must be more precise than just 'to make it safe to drink'.

.... **(1 mark)**

Guided

(b) Drinking water contains low concentrations of dissolved chlorine. Suggest reasons that explain why this water is considered safe to drink, even though chlorine gas is toxic.

The concentration of chlorine is high enough to ....

but low enough so that it .... **(2 marks)**

3 Name the **two** stages in water treatment that are carried out before chlorine is added. Give a reason why each stage is carried out.

> You do not have to place the two stages in a correct order in this question.

name of stage: ....

reason: ....

name of stage: ....

reason: .... **(4 marks)**

4 Explain why distilled water, rather than tap water, is used in chemical analysis.

Guided

Unlike tap water, distilled water does not contain ....

These would .... **(2 marks)**

5 Drinking water in the UK comes from fresh water including rivers, lakes and reservoirs. In some countries drinking water may come from sea water instead.

(a) Name a separation method used to separate water for drinking from sea water.

.... **(1 mark)**

(b) Suggest a reason that explains why producing drinking water from sea water is usually expensive.

.... **(1 mark)**

6 Aluminium sulfate may be added during water treatment. It forms a precipitate of aluminium hydroxide, which traps small solid particles suspended in the water. Balance the equation for this reaction.

$Al_2(SO_4)_3(aq) + ...H_2O(l) \rightarrow ....Al(OH)_3(s) + ...H_2SO_4(aq)$ **(1 mark)**

Extended response – Separating mixtures

A cloudy pale-yellow mixture contains three substances (**A**, **B** and **C**).

The table shows some information about these substances.

Substance	Melting point (°C)	Boiling point (°C)	Notes
A	115	445	yellow, insoluble in **B** and **C**
B	−95	56	colourless, soluble in **C**
C	0	100	colourless, soluble in **B**

Devise a method to produce pure samples of each individual substance from the mixture.

> The command word **devise** means that you are being asked to plan or invent a procedure from existing principles or ideas. You do not have to imagine a complex method that goes beyond your GCSE studies.

You should use the information in the table in your answer, and explain why you have suggested each step.

> The separation methods covered at GCSE include simple distillation, fractional distillation, filtration, crystallisation and paper chromatography. You do not need to use them all to answer this question.

> Work out whether each substance is a solid, liquid or gas. Also work out whether they mix to form a solution. Then think about the properties that could be used to separate each substance from the others. How could you separate **A** from **B** and **C**, and **B** from **C**?

....

....

....

....

....

....

....

....

....

....

.... **(6 marks)**

> You should be able to describe an appropriate experimental technique to separate a mixture if you know the properties of the components of the mixture.

Acids and alkalis

1 Acids and alkalis in solution are sources of ions. Which of the following is correct?

☐ **A** Acids are a source of hydroxide ions, H^+.

☐ **B** Acids are a source of hydrogen ions, OH^-.

☐ **C** Alkalis are a source of hydrogen ions, H^+.

☐ **D** Alkalis are a source of hydroxide ions, OH^-.   **(1 mark)**

2 A student adds a few drops of universal indicator solution to some dilute hydrochloric acid.

(a) The universal indicator solution is green before it is added to the acid. Explain
what this tells you about the pH of the universal indicator solution.

....

.... **(2 marks)**

(b) Give the colour of the mixture formed by the universal indicator solution and
dilute hydrochloric acid.

.... **(1 mark)**

3 A student heats some magnesium ribbon in air. It burns with a white flame and a
white solid forms. The student then mixes the white solid with water in a test tube.

(a) Balance the equation for the reaction
between magnesium and oxygen.

> The equation is balanced when the number of atoms of each element is the same on both sides.

........$Mg + O_2 \rightarrow$ ....MgO **(1 mark)**

(b) Universal indicator solution turns purple when it is added to the mixture in the
test tube. State what this tells you about the mixture.

.... **(1 mark)**

4 Complete the table by placing
a tick (✓) in each correct box to
show whether a substance is an
acid or an alkali.

> You should be able to recall the formulae of elements and simple compounds.

Formula of substance	Type of substance	
	Acid	Alkali
NaOH		
HCl		
H_2SO_4		
NH_3		

(4 marks)

5 (a) Complete the table to show
the colours of litmus, methyl
orange and phenolphthalein at
different pH values.

> You should be able to recall the effect of acids and alkalis on these indicators.

Indicator	Colour at pH 14	Colour at pH 1
litmus		
methyl orange	yellow	red
phenolphthalein		

(3 marks)

(b) Predict the colour of litmus solution in pure water.

.... **(1 mark)**

Bases and alkalis

1   What forms when an acid reacts with a metal hydroxide?

☐ **A** a salt only

☐ **C** a salt and hydrogen only

☐ **B** a salt and water only

☐ **D** a salt and carbon dioxide only   **(1 mark)**

2   Sodium carbonate reacts with dilute nitric acid, forming three products.

(a) Name the salt formed in this reaction.

....   **(1 mark)**

(b) Write a word equation for this reaction.

....   **(1 mark)**

(c) Describe **two** things that you would see when sodium carbonate powder is added to dilute nitric acid.

> You do not need to name any substances in your answer.

1: ....

2: ....   **(2 marks)**

3   Calcium reacts with dilute hydrochloric acid. Bubbles of gas are given off and a colourless solution forms.

(a) Name the colourless solution that forms.

....   **(1 mark)**

(b) Name the gas responsible for the bubbles.

....   **(1 mark)**

4   Describe the chemical test for:

> **Practical skills** Write down what you would do and what you would expect to observe.

(a) carbon dioxide

....

....   **(2 marks)**

(b) hydrogen

....

....   **(2 marks)**

5   Zinc oxide is an example of a base.

> **Guided**

(a) Describe what is meant by a **base**.

A base is any substance that reacts with ....

to form a .... and .... only.   **(2 marks)**

(b) State the general name for a **soluble** base.

....   **(1 mark)**

(c) Name the salt formed when zinc oxide reacts with dilute sulfuric acid.

....   **(1 mark)**

Neutralisation

1 Acids react with alkalis to form salts and water only. Explain, in terms of reacting ions, how water forms in these neutralisation reactions.

Guided

Hydrogen ions, H^+, from the .... react with

.... ions from the ....

to form ....   **(3 marks)**

2 Dilute hydrochloric acid, HCl, reacts with calcium oxide and calcium hydroxide.

(a) Balance the equation for the reaction of calcium oxide with dilute hydrochloric acid.

$CaO + ...HCl \rightarrow CaCl_2 + H_2O$ **(1 mark)**

(b) Write a balanced equation for the reaction of calcium hydroxide, $Ca(OH)_2$, with dilute hydrochloric acid.

> The substances formed are the same as those in the reaction between calcium oxide and hydrochloric acid.

....   **(2 marks)**

> You should be able to use appropriate apparatus and substances to measure the pH in different situations.

3 Limewater is calcium hydroxide solution. A student investigates what happens to the pH when he adds small portions of limewater to 25 cm³ of dilute hydrochloric acid in a flask. The table shows his results.

Volume of limewater added (cm³)	pH of the mixture in the flask
0	1.6
5	1.8
10	2.0
15	2.2
20	2.6
24	3.8
25	7.0
26	10.4
30	11.2
35	11.5
40	11.6

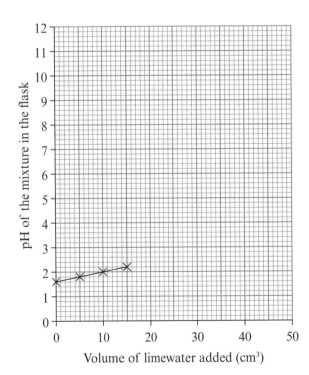

Complete the graph by plotting the remaining results.   **(2 marks)**

> **Maths skills** If you are asked to plot a graph, you need to mark the points accurately on the grid, then draw a line or curve of best fit. You must also work out a suitable scale and label the axes if this has not already been done for you.

Salts from insoluble bases

1 Iron(II) oxide is an insoluble base. It reacts with dilute sulfuric acid to form iron(II) sulfate solution and water:

$$FeO(s) + H_2SO_4(aq) \rightarrow FeSO_4(aq) + H_2O(l)$$

(a) Explain why an excess of iron(II) oxide is added to the dilute sulfuric acid.

> If a reactant is in excess, some of it is left over when the reaction has finished.

....

.... **(2 marks)**

(b) Suggest a reason that explains why the dilute sulfuric acid may be warmed before adding iron(II) oxide.

> What factors influence the rate of a reaction?

.... **(1 mark)**

(c) Name the separation method needed to remove the excess iron(II) oxide.

.... **(1 mark)**

(d) Name the process used to produce iron(II) sulfate crystals from the iron(II) sulfate solution.

.... **(1 mark)**

2 A student wants to prepare pure, dry crystals of copper sulfate. This is her method.

> **Making copper sulfate crystals**
>
> **A** Put 25 cm³ of dilute sulfuric acid in a beaker.
> **B** Add several spatulas of copper oxide powder.
> **C** Pour the liquid from the beaker into an evaporating basin.
> **D** Heat the liquid using a blue Bunsen burner with the air hole fully open until all the water has boiled away.

(a) Name a suitable piece of apparatus to measure 25 cm³ of dilute sulfuric acid at step A.

.... **(1 mark)**

(b) Describe **two** improvements that the student could make at step B, and give reasons for your answers.

improvement 1: ....

reason: ....

improvement 2: ....

reason: .... **(4 marks)**

(c) Describe **one** improvement that the student could make at step C, and give a reason for your answer.

> Think about whether there are any substances contaminating the copper sulfate solution.

improvement: ....

reason: .... **(2 marks)**

116

Salts from soluble bases

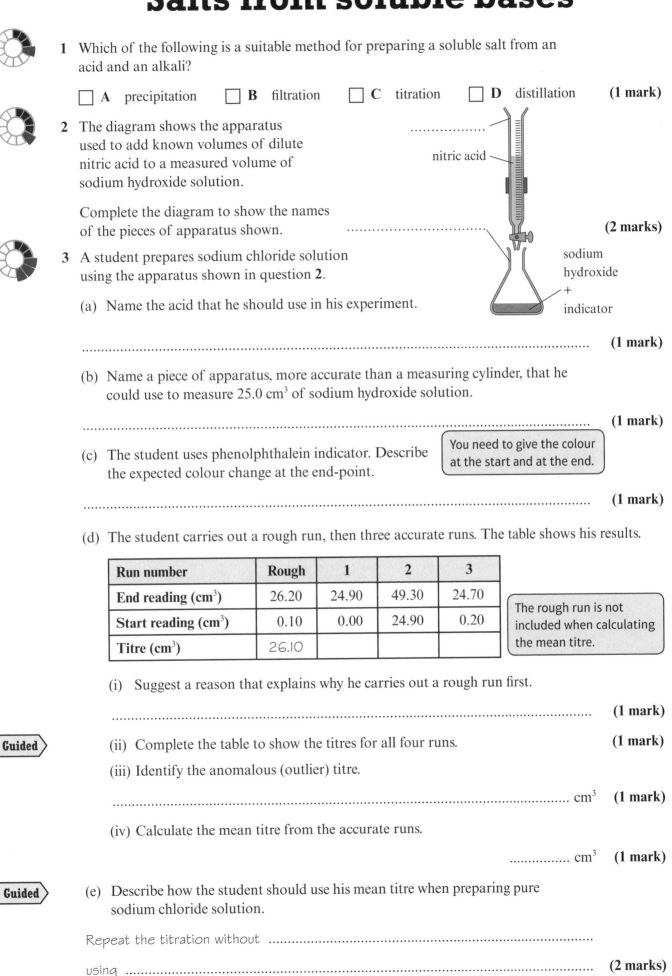

1 Which of the following is a suitable method for preparing a soluble salt from an acid and an alkali?

☐ **A** precipitation   ☐ **B** filtration   ☐ **C** titration   ☐ **D** distillation   **(1 mark)**

2 The diagram shows the apparatus used to add known volumes of dilute nitric acid to a measured volume of sodium hydroxide solution.

....

nitric acid

Complete the diagram to show the names of the pieces of apparatus shown.   **(2 marks)**

....

3 A student prepares sodium chloride solution using the apparatus shown in question **2**.

sodium hydroxide + indicator

(a) Name the acid that he should use in his experiment.

.... **(1 mark)**

(b) Name a piece of apparatus, more accurate than a measuring cylinder, that he could use to measure 25.0 cm³ of sodium hydroxide solution.

.... **(1 mark)**

(c) The student uses phenolphthalein indicator. Describe the expected colour change at the end-point.

> You need to give the colour at the start and at the end.

.... **(1 mark)**

(d) The student carries out a rough run, then three accurate runs. The table shows his results.

Run number	Rough	1	2	3
End reading (cm³)	26.20	24.90	49.30	24.70
Start reading (cm³)	0.10	0.00	24.90	0.20
Titre (cm³)	26.10			

> The rough run is not included when calculating the mean titre.

(i) Suggest a reason that explains why he carries out a rough run first.

.... **(1 mark)**

(ii) Complete the table to show the titres for all four runs.   **(1 mark)**

(iii) Identify the anomalous (outlier) titre.

.... cm³   **(1 mark)**

(iv) Calculate the mean titre from the accurate runs.

.... cm³   **(1 mark)**

(e) Describe how the student should use his mean titre when preparing pure sodium chloride solution.

Repeat the titration without ....

using .... **(2 marks)**

Making insoluble salts

1 Potassium chloride is a metal chloride that is soluble in water. Which of the following metal chlorides is insoluble in water?

> You need to be able to recall the general rules for the solubility of common types of substances in water.

☐ **A** sodium chloride

☐ **B** silver chloride

☐ **C** copper chloride

☐ **D** zinc chloride   **(1 mark)**

2 Which of the following pairs contains one substance that is soluble in water and one that is insoluble in water?

☐ **A** lead chloride and barium sulfate

☐ **B** calcium nitrate and potassium hydroxide

☐ **C** aluminium hydroxide and copper carbonate

☐ **D** ammonium carbonate and calcium sulfate   **(1 mark)**

3 A student wants to produce insoluble calcium hydroxide.

(a) Name two solutions that, when mixed together, will produce a precipitate of calcium hydroxide.

> To make a precipitate XY, you can mix X nitrate with sodium Y.

solution 1: ....

solution 2: ....   **(2 marks)**

(b) Name the other product formed when the two solutions named in (a) are mixed together.

....   **(1 mark)**

(c) Suggest a reason that explains why the student does not need to heat the solutions first.

> **Practical skills** Think about how quickly precipitation reactions happen, and what heating does to the rate of a reaction.

....   **(1 mark)**

4 Sodium carbonate, Na_2CO_3, and calcium chloride, $CaCl_2$, are soluble in water. Calcium carbonate, $CaCO_3$, is insoluble in water.

(a) Balance the equation for the reaction between sodium carbonate solution and calcium chloride solution. Include state symbols in your answer.

Na_2CO_3(......) + $CaCl_2$(......) → ...... $NaCl$(......) + $CaCO_3$(......)   **(2 marks)**

> **Guided**

(b) Describe how you would use solid sodium carbonate and solid calcium chloride to produce a pure, dry sample of calcium carbonate.

Add water to .... then mix.

Separate the precipitate of calcium carbonate using ....

Wash the .... using ....

then dry it by ....   **(4 marks)**

Extended response – Making salts

Sodium chloride solution can be made from dilute hydrochloric acid and sodium hydroxide solution:

$$HCl(aq) + NaOH(aq) \rightarrow NaCl(aq) + H_2O(l)$$

Devise a titration experiment to find the exact volume of hydrochloric acid needed to neutralise 25.0 cm³ of sodium hydroxide solution. Explain how you would use the result from this experiment to obtain pure, dry, sodium chloride crystals.

> It may help to plan your answers to questions similar to this one. For example, you could divide your answer here into three sections:
>
> - setting up the apparatus ready for a titration, including where the reagents need to go
> - carrying out the titration, including steps needed to obtain an accurate result
> - producing sodium chloride crystals from sodium chloride solution.

....

....

....

....

....

....

....

....

....

....

....

....

....

....

....

....

....

.... **(6 marks)**

> You should be able to describe how to carry out an acid–alkali titration using a burette, a pipette and a suitable indicator, to prepare a pure, dry salt.

Electrolysis

1   Under what conditions can an ionic compound conduct electricity?

> Molten substances are in the liquid state.

☐ **A**   only when it is molten

☐ **B**   when it is solid or molten

☐ **C**   when it is solid or dissolved in water

☐ **D**   when it is molten or dissolved in water **(1 mark)**

2   Complete the table by placing a tick (✓) in each correct box to describe some features of electrolysis.

> You will need one tick in each row.

	Positively charged	Negatively charged
Anode		
Anion		
Cathode		
Cation		

(4 marks)

3   State what is meant by the term '**electrolyte**'.

> Guided

An electrolyte is an .... compound in the .... state

or .... **(2 marks)**

4   In an electrolysis experiment, molten zinc bromide is decomposed.

Zinc forms at the cathode. Predict the product that forms at the anode.

.... **(1 mark)**

5   Sodium is extracted from molten sodium chloride, NaCl, by electrolysis. The reaction at one of the electrodes can be modelled as: $Na^+ + e^- \rightarrow Na$

State at which electrode, the positively or negatively charged electrode, this reaction happens. Give a reason for your answer using information from the equation.

> Look at the charge on the sodium ions.

....

.... **(1 mark)**

6   A student places a purple crystal of potassium manganate(VII), $KMnO_4$, on a damp piece of filter paper. She connects each end of the paper to a DC electricity supply.
A purple streak gradually moves to the left.

(+   potassium manganate(VII)   −)

> Potassium manganate(VII) contains manganate(VII) ions, MnO_4^-, and colourless potassium ions, K^+.

Explain why the purple streak moves to the left.

....

.... **(2 marks)**

Electrolysing solutions

1 The ions in copper chloride solution are:

- copper ions, Cu^{2+}
- chloride ions, Cl^-
- hydrogen ions, H^+
- hydroxide ions, OH^-

Copper chloride solution is electrolysed using a DC electricity supply.

(a) Which of these ions form from the water in the copper chloride solution?

☐ **A** H^+ and Cu^{2+} ions

☐ **B** H^+ and OH^- ions

☐ **C** H^+ ions only

☐ **D** OH^- ions only   **(1 mark)**

(b) Which of these ions will be attracted to the cathode during the electrolysis of copper chloride solution?

> The cathode is the negatively charged electrode.

☐ **A** Cl^- ions only

☐ **B** Cl^- ions and OH^- ions

☐ **C** H^+ ions only

☐ **D** H^+ and Cu^{2+} ions   **(1 mark)**

(c) Predict the substance formed at the cathode during the electrolysis of concentrated copper chloride solution.

.... **(1 mark)**

2 The electrolysis of concentrated sodium chloride solution, $NaCl(aq)$, produces two useful gases.

(a) Write the formulae of all the ions present in a concentrated sodium chloride solution.

> You should be able to recall the formulae of ions.

.... **(2 marks)**

(b) Predict the gas that forms at:

(i) the anode .... **(1 mark)**

(ii) the cathode .... **(1 mark)**

3 Water, acidified with sulfuric acid, can be decomposed by electrolysis. Complete the table by writing the correct product into each box.

Electrode	Product formed
anode	
cathode	

(2 marks)

4 Oxygen is produced at the anode during the electrolysis of sodium sulfate solution.

Name, or give the formula of, the ions present in this solution that are discharged to form oxygen.

> Which negatively charged ions will be present in this solution?

.... **(1 mark)**

 **Investigating electrolysis**

A student researched the different types of electrode used for electrolysis. This is what he found out.

> **Electrodes**
>
> Electrodes can be inert or non-inert. Graphite electrodes are inert electrodes – they just provide a surface for electrode reactions to happen. Copper electrodes are non-inert electrodes for the electrolysis of copper sulfate solution. Their copper atoms may form copper ions, causing the electrode to lose mass.

1 Oxygen forms during the electrolysis of copper sulfate solution using graphite electrodes.

> Look at the information that the student discovered.

(a) State whether the electrodes are inert or non-inert in this electrolysis reaction.

.... **(1 mark)**

(b) Explain at which electrode (anode or cathode) oxygen will be produced.

....

.... **(2 marks)**

2 The electrolysis of copper sulfate solution, using copper electrodes, is used to purify copper. During electrolysis, the copper anode loses mass. The copper cathode gains mass because copper is deposited.

A student investigates the gain in mass by a copper cathode. She runs each experiment for the same time, but changes the current. She measures the mass of the cathode before and after electrolysis. The graph shows her results.

(a) Identify the variable controlled by the student in her experiment.

> Variables are factors that can be measured or observed.

.... **(1 mark)**

(b) Identify the dependent variable in the student's experiment.

.... **(1 mark)**

(c) State whether the electrodes are inert or non-inert in this experiment, and justify your answer.

....

.... **(1 mark)**

> **Guided**

(d) Calculate the gradient of the line of best fit. Give your answer to two significant figures.

change on the y-axis = 0.15 – 0.04 = ...

change on the x-axis = ... – .... = ...

gradient = ...

= .... g/A

> 🖩 **Maths skills** A linear relationship such as this can be represented by: $y = mx + c$ m is the gradient and c is the intercept on the vertical axis (y-axis). The gradient equals the change on the y-axis, divided by the change on the x-axis.

(3 marks)

122

Extended response – Electrolysis

A student carries out two experiments using copper chloride, $CuCl_2$.

In experiment 1, the student places two graphite electrodes into solid copper chloride powder in a beaker. She then connects the electrodes to a DC electricity supply and records any changes.

For experiment 2, the student disconnects the DC supply, then adds some water to dissolve the copper chloride. She reconnects the electrodes to the DC supply and records any changes.

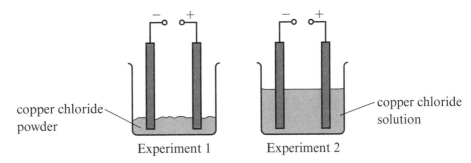

copper chloride powder

copper chloride solution

Experiment 1   Experiment 2

The table shows the student's results.

Experiment	Observations at the cathode (–)	Observations at the anode (+)
1	no visible change	no visible change
2	brown solid forms on the electrode	bubbles of a pale yellow–green gas released

Explain the differences between the results shown in the table for experiments 1 and 2. You should include a balanced equation for the overall reaction in your answer.

> Explain why copper chloride powder does not conduct electricity, and then explain why copper chloride solution does conduct electricity. Name the substances responsible for the student's observations at each electrode during experiment 2.

....

....

....

....

....

....

....

....

....

.... **(6 marks)**

> Explain, in terms of the ions in the copper chloride solution, why each substance forms.

The reactivity series

1   Four metals (**W**, **X**, **Y** and **Z**) are added to cold water and to dilute hydrochloric acid.
The table shows what happens.

Metal	Observations in water	Observations in dilute hydrochloric acid
W	slow bubbling	very fast bubbling
X	no visible change	no visible change
Y	fast bubbling	very fast bubbling
Z	no visible change	slow bubbling

(a) Which of the following shows the order of reactivity, from most reactive to least reactive metal?

☐ **A** W, Y, X, Z

☐ **B** X, Z, W, Y

☐ **C** Y, W, X, Z

☐ **D** Y, W, Z, X   **(1 mark)**

(b) The concentration of hydrochloric acid is kept the same each time. Give two other variables that should be kept the same in each experiment so that the reactivity of the metals can be compared.

....

.... **(2 marks)**

2   Name **one** common metal that does not react with dilute acids.

> Metals that are less reactive than hydrogen do not react with dilute acids.

.... **(1 mark)**

3   Magnesium reacts slowly with cold water to produce magnesium hydroxide, $Mg(OH)_2$, and a gas.

> This flammable gas is produced when any metal reacts with water or dilute acids.

(a) (i) Name the gas produced in the reaction.

.... **(1 mark)**

   (ii) Describe the chemical test used to identify the gas produced in the reaction.

....

.... **(2 marks)**

(b) Name the compound formed when magnesium reacts with steam, rather than cold water.

.... **(1 mark)**

4   Aluminium is protected from contact with water by a natural layer of aluminium oxide, Al_2O_3. This means that aluminium does not react with water, even though it is a reactive metal. However, aluminium does react with dilute acids, such as dilute sulfuric acid.

(a) Balance the equation for the reaction between aluminium oxide and dilute sulfuric acid.

$Al_2O_3(s) + ...H_2SO_4(aq) \rightarrow Al_2(SO_4)_3(aq) + ...H_2O(l)$ **(1 mark)**

> Guided

(b) There is no immediate visible change when aluminium is added to dilute sulfuric acid. Bubbling then starts and gets increasingly fast. Suggest reasons that explain these observations.

To begin with, the acid reacts with ....

but, once this has gone, it reacts with .... **(2 marks)**

Metal displacement reactions

1 Copper can displace silver from silver nitrate solution. Copper nitrate solution also forms in the reaction.

> Which metal, copper or silver, is the more reactive of the two metals?

(a) Give a reason to explain why copper can displace silver from silver salts in solution.

.... **(1 mark)**

 (b) Balance the equation for the reaction, and include state symbols.

Cu(........) + ... AgNO₃(aq) → ....Ag(s) + Cu(NO₃)₂(........)   **(2 marks)**

2 A student investigates the reactivities of four metals, copper, magnesium, zinc and X. She adds pieces of magnesium ribbon to solutions of the nitrates of each metal. She then removes and examines each piece of magnesium ribbon after a few minutes. The table shows her results.

Experiment	Solution used	Observations
1	copper nitrate	brown coating on the magnesium ribbon
2	magnesium nitrate	no visible change
3	zinc nitrate	grey coating on the magnesium ribbon
4	X nitrate	grey coating on the magnesium ribbon

(a) Name the substance found in the brown coating on the magnesium ribbon in Experiment 1.

.... **(1 mark)**

(b) State why there is no visible change in Experiment 2.

.... **(1 mark)**

(c) The student repeats the experiment with metal X instead of magnesium.

Experiment	Solution used	Observations
5	copper nitrate	brown coating on the piece of metal X
6	magnesium nitrate	no visible change
7	zinc nitrate	grey coating on the piece of metal X
8	X nitrate	no visible change

(i) Use the results shown in both tables to place the **two most reactive** metals in order of **decreasing** reactivity.

....

.... **(1 mark)**

(ii) Describe **one** experiment needed to find the order of reactivity of the other two metals.

....

.... **(2 marks)**

3 The thermite reaction makes molten iron for welding railway lines:

2Al(s) + Fe₂O₃(s) → Al₂O₃(s) + 2Fe(l)

Explain what this reaction shows about the relative reactivity of aluminium and iron.

....

.... **(2 marks)**

Explaining metal reactivity

1   Give the meaning of the term 'cation'.

> **Guided**

A cation is a .... charged ion.   **(1 mark)**

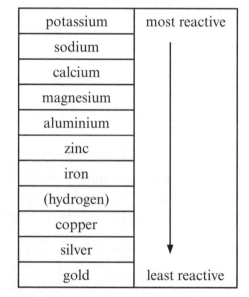

2   Calcium is a reactive metal. It reacts vigorously with dilute hydrochloric acid to form calcium chloride solution and hydrogen gas:

$$Ca(s) + 2HCl(aq) \rightarrow CaCl_2(aq) + H_2(g)$$

(a)   The formula for a chloride ion is Cl^-.
Deduce the formula for a calcium ion.

> Remember that the overall charge in $CaCl_2$ will be 0.

....   **(1 mark)**

> **Guided**

(b)   Describe what happens when a calcium atom becomes a calcium ion.

> Mention how many electrons are lost and from where.

.... electrons are lost from the ....   **(2 marks)**

(c)   The table shows a reactivity series for the metals. Hydrogen is a non-metal. It is included for comparison.

> The more easily a metal's atoms form cations, the more reactive the metal is.

Identify the metal that:

(i)   forms cations most easily   **(1 mark)**

....

(ii)   forms cations least easily.

....
(1 mark)

potassium	most reactive
sodium	
calcium	
magnesium	
aluminium	
zinc	
iron	
(hydrogen)	
copper	
silver	
gold	least reactive

(d)   Identify a metal that will **not** react with dilute acids.

> Look at the position of hydrogen (a non-metal) in the reactivity series.

....   **(1 mark)**

3   Zinc displaces copper from copper sulfate solution:

$$Zn(s) + CuSO_4(aq) \rightarrow ZnSO_4(aq) + Cu(s)$$

Explain this reaction in terms of the tendency to form cations.

....

....

....   **(2 marks)**

Metal ores

1   Some metals are found in the Earth's crust as uncombined metals. Which of these metals is found uncombined in the Earth's crust?

    ☐ **A**   potassium   ☐ **C**   gold

    ☐ **B**   zinc   ☐ **D**   aluminium   **(1 mark)**

2   Tungsten metal is extracted by heating tungsten oxide, WO_3, in a stream of hydrogen gas, H_2:

$$WO_3 + 3H_2 \rightarrow W + 3H_2O$$

    (a)   What happens in this reaction?

       ☐ **A**   Tungsten oxide is reduced.   ☐ **C**   Tungsten is oxidised.

       ☐ **B**   Hydrogen is reduced.   ☐ **D**   Water is oxidised.   **(1 mark)**

    (b)   Suggest a reason that explains why this process is hazardous.

> Look at the reactants and products – could any of them cause harm to people or objects?

    ....   **(1 mark)**

3   Metals are extracted from their ores. Give the meaning of the term '**ore**'.

> **Guided**

    a rock or mineral that contains enough ....

    to make its extraction ....   **(2 marks)**

4   Tin is produced when tin oxide is heated with carbon.

    (a)   Complete the word equation for the reaction.

    tin oxide + carbon → .... + ...   **(2 marks)**

    (b)   Explain whether carbon is oxidised or reduced in this reaction.

> Think about the gain or loss of oxygen in the reaction.

    ....

    ....   **(2 marks)**

5   Corrosion occurs when a metal oxidises, and this process continues. For example, sodium is shiny when it is freshly cut, but a dull layer of sodium oxide forms quickly when sodium is exposed to air.

    (a)   Suggest a reason that explains why sodium metal is stored in oil.

    ....   **(1 mark)**

    (b)   Give a reason that explains why copper oxidises slowly unless it is heated strongly.

    ....

    ....   **(1 mark)**

Iron and aluminium

1   The table shows a reactivity series for the metals. Name a metal in the table that:

(a) is more reactive than carbon.

.... **(1 mark)**

(b) could be extracted from its ore by heating with carbon.

.... **(1 mark)**

sodium	most reactive
calcium	
magnesium	
(carbon)	
lead	
copper	least reactive

2   Metals are extracted from their ores in different ways. They may be extracted either by heating with carbon or by electrolysis. Iron is extracted from iron oxide by heating with carbon.

(a) Complete this word equation for the extraction of iron.

iron oxide + carbon → .... + ...   **(2 marks)**

(b) State why iron can be extracted from iron oxide by heating with carbon.

> Think about the reactivity series.

.... **(1 mark)**

(c) State whether iron oxide is oxidised or reduced in this reaction, and give a reason for your answer.

.... **(1 mark)**

3   Bauxite is an aluminium ore. It contains aluminium oxide, Al_2O_3. Aluminium is extracted from purified aluminium oxide by electrolysis.

(a) Give a reason why electrolysis is more expensive than heating with carbon.

> Think about the amount of energy involved in each method of extraction.

.... **(1 mark)**

(b) Predict the products formed at each electrode during the electrolysis of aluminium oxide.

> The cathode is the negatively charged electrode.

(i)   at the cathode: .... **(1 mark)**

(ii) at the anode: .... **(1 mark)**

(c) The anodes are made from graphite, a form of carbon. Suggest a reason that explains why the anodes must be replaced frequently.

.... **(1 mark)**

4   Zinc could be extracted either by heating zinc oxide with carbon or by electrolysis of molten zinc oxide. Explain which method is most likely to be used.

> The method chosen is related to the position of a metal in the reactivity series and the cost of extracting it.

Guided

Zinc is most likely to be extracted by ....

....

because ....

.... **(2 marks)**

Recycling metals

1 Complete the table by placing a tick (✓) in the box against each **disadvantage** of recycling metals.

Feature of recycling metals	Disadvantage (✓)
Used metal items must be collected.	
The use of finite resources is decreased.	
Different metals must be sorted.	
Metals can be melted down.	

Do not place a cross against the incorrect statement(s) – you are asked to place only ticks in the table.

(2 marks)

2 Metal ores are removed from the ground in large amounts by quarrying. This may involve using explosives to break up the rock, and large machinery and vehicles to carry the rock away. Give **two** ways in which quarrying can damage the local environment.

1 ....

2 .... **(2 marks)**

Guided

3 Around 90% of the lead produced each year is used in traditional 'lead acid' batteries for cars and other vehicles. About 70% of the lead used each year is recycled lead.

(a) Describe an advantage of recycling lead from lead acid batteries, rather than recycling lead from general scrap metal waste.

Think about how different metals are obtained from scrap metal waste.

Most lead for recycling is found in ....

so lead does not need to be .... **(2 marks)**

(b) Describe **two** advantages of recycling metals, rather than extracting them from their compounds.

1 ....

2 .... **(2 marks)**

4 Some food cans are made from aluminium. Others are made from steel (an iron alloy) coated with tin. The table shows some information about these three metals.

Metal	Abundance in the Earth's crust (%)	Cost of 1 tonne of metal (£)	Energy saved by recycling (%)
iron	6.3	500	70
tin	0.00022	16 500	75
aluminium	8.2	1500	94

(a) Identify the metal for which the most energy is saved by recycling.

Use information from the table.

.... **(1 mark)**

(b) Give **two** reasons why it may be more important to recycle tin, rather than iron or aluminium. Use information from the table.

1 ....

2 .... **(2 marks)**

Life-cycle assessments

Guided

1 A life-cycle assessment for a manufactured product involves considering its effect on the environment at all stages.

The table shows the main steps involved in a life-cycle assessment. Give the correct order by writing the numbers 1 to 4 in the correct boxes, where step 1 is the first.

Step number	Process
	manufacturing the product
	obtaining raw materials
4	disposing of the product
	using the product

(1 mark)

Guided

2 Manufacturers can make glass bottles with thinner walls than in the past. A bottle for a fizzy drink had a mass of 0.24 kg in 1996 but a mass of 0.20 kg in 2016.

(a) 16.5 MJ/kg of glass is used in their manufacture.

   (i) Calculate the energy, in MJ, needed to make one bottle in 1996.

   energy = 16.5 × 0.24

   = ... MJ **(1 mark)**

   (ii) Calculate the energy, in MJ, needed to make one bottle in 2016.

   .... MJ **(1 mark)**

(b) Carbon dioxide, CO_2, is a greenhouse gas. The manufacture of glass bottles causes the emission of 1.2 kg of CO_2/kg glass. Calculate the difference, between 1996 and 2016, in the mass of carbon dioxide emitted when one bottle is made.

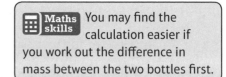
Maths skills You may find the calculation easier if you work out the difference in mass between the two bottles first.

difference in mass between bottles in 1996 and 2016 = 0.24 − 0.20

= .... kg

difference in the mass of CO_2 emitted = ... kg **(2 marks)**

3 Window frames may be made from either PVC (a polymer) or wood. The table shows some information from a life-cycle assessment of a window frame.

Process	Energy used (MJ)	
	PVC frame	Wooden frame
producing the material	12.0	4.0
making the frame	3.0	3.6
transport and installation	4.2	4.8
maintenance	0.3	1.5
disposal in landfill	0.7	0.8

(a) Identify the stage in the life cycle of each window frame that is responsible for the most energy use.

PVC frame: ....

wooden frame: .... **(1 mark)**

(b) Explain, using data in the table, which type of frame is likely to have the lower environmental impact **when in use**.

....

.... **(2 marks)**

The Haber process

1 In the Haber process, nitrogen and hydrogen react together to form ammonia:

$$N_2(g) + 3H_2(g) \rightleftharpoons 2NH_3(g)$$

	Nitrogen	Hydrogen
☐ A	air	hydrochloric acid
☐ B	natural gas	air
☐ C	sea water	natural gas
☐ D	air	natural gas

(a) Which row correctly shows the raw materials for nitrogen and hydrogen?

(1 mark)

(b) Give the meaning of the symbol $\rightleftharpoons$ in the balanced equation.

.... **(1 mark)**

2 The conditions used in the Haber process are carefully controlled to achieve an acceptable yield of ammonia in an acceptable time.

(a) State the temperature and pressure used in the Haber process.

temperature: .... °C

pressure: .... atmospheres **(2 marks)**

(b) Explain why iron is added to the reactor used in the Haber process.

> Iron affects the reaction but is not a reactant or a product.

....

.... **(2 marks)**

3 Dilute ethanoic acid reacts with ethanol. Ethyl ethanoate and water form in the reaction:

> You may see familiar chemistry in an unfamiliar context, like this one.

ethanoic acid + ethanol $\rightleftharpoons$ ethyl ethanoate + water

All four substances are clear, colourless liquids. They mix completely with each other.

> Do not be put off by the complex appearance of the equation. It is just an example of: A + B $\rightleftharpoons$ C + D

(a) State what visible changes, if any, you would observe during the reaction.

.... **(1 mark)**

Guided (b) The reaction reaches a dynamic equilibrium after a few days.

  (i) Describe what is happening to the forward and backward reactions at equilibrium.

  The rate of the forward and backward reactions is ....

  and they .... **(2 marks)**

  (ii) State what happens to the concentrations of the reacting substances at equilibrium.

  > The choices you have are: increase, decrease, do not change.

.... **(1 mark)**

Extended response – Reactivity of metals

Magnesium forms cations more readily than copper. A spatula of magnesium powder mixed with a spatula of copper oxide powder is heated strongly on a steel lid. Magnesium oxide and copper are produced:

magnesium + copper oxide → magnesium oxide + copper
$$Mg(s) + CuO(s) \rightarrow MgO(s) + Cu(s)$$

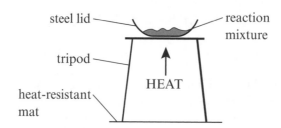

The reaction shows that magnesium is more reactive than copper. It is a very vigorous reaction. Energy is transferred to the surroundings by light and heating, and hot powder escapes into the air.

Devise an experiment, based on this method, to investigate the relative reactivity of copper, iron and zinc. In your answer, describe the results that you expect, and explain how you would use them to deduce the order of reactivity. Explain how you would control the risks of harm in the investigation.

> Think about how many combinations of a metal powder and a metal oxide powder you will need to test.
>
> One way to show these combinations, and the expected results, is to make a completed results table.

....

....

....

....

....

....

....

....

....

....

.... **(6 marks)**

> You should be able to evaluate the risks in a practical procedure, and suggest suitable precautions for a range of practicals (not just those mentioned in the specification).

The alkali metals

1 Compared with a typical transition metal such as iron, the alkali metals are:

☐ **A** hard with relatively low melting points

☐ **B** soft with relatively high melting points

☐ **C** soft with relatively low melting points

☐ **D** hard with relatively high melting points   **(1 mark)**

2 Complete the table to describe the reactions of lithium, sodium and potassium with water.

Alkali metal	Flame colour	Description
lithium	does not ignite	fizzes steadily disappears slowly
sodium	orange if ignited	
potassium		

(5 marks)

3 The alkali metals react with water to produce a metal hydroxide and hydrogen. For example:

   sodium + water → sodium hydroxide + hydrogen

> You will need an even number of hydrogen atoms on each side.

Balance the equation for this reaction. Include state symbols.

......Na(......) + ...H_2O(......) → ......NaOH(aq) + H_2(......)   **(2 marks)**

4 Give a reason that explains why, in terms of electronic configurations, the alkali metals occupy Group 1.

> The electronic configurations of the atoms of these elements differ, but they do have something in common.

.... **(1 mark)**

5 Explain why lithium, sodium and potassium are stored in oil.

....

.... **(2 marks)**

6 Francium, Fr, is placed at the bottom of Group 1.

> The Group 1 hydroxides all have similar formulae.

(a) Predict the formula of francium hydroxide.

.... **(1 mark)**

(b) Predict one observation that you would expect to see in the reaction of francium with water.

.... **(1 mark)**

7 Reactivity increases going down Group 1. Explain this reactivity pattern in terms of the electronic configurations of the atoms.

Guided

Going down the Group, the size of the atoms...

The outer electron becomes ....

so the outer electron is lost .... **(3 marks)**

The halogens

1   Which of the following is a chemical test for chlorine gas?

☐ **A**   Damp red litmus paper turns blue, then white.

☐ **B**   Damp blue litmus paper turns red, then white.

☐ **C**   Damp starch–iodide paper turns red, then white.

☐ **D**   Dry starch–iodide paper turns blue–black.   **(1 mark)**

> **Practical skills**   Answer D cannot be correct because chlorine must dissolve in water for the chemical test to work.

2   Give a reason that explains why, in terms of electronic configurations, the halogens occupy Group 7.

....   **(1 mark)**

3   (a)   Complete the table to show the colours and physical states of the halogens at room temperature.

> You must be able to recall the colours and states of chlorine, bromine and iodine at room temperature.

Halogen	Colour	Physical state
chlorine		
bromine		
iodine	dark grey	solid (forms a purple vapour)

(4 marks)

(b)   Astatine is the element placed immediately below iodine. Predict its colour and physical state.

....   **(2 marks)**

4   The table shows the densities of two halogens, in order going down Group 7.

(a)   Predict the density of astatine, the element placed immediately below iodine. Write its predicted density into the table.   **(1 mark)**

Halogen	Density at room temperature and pressure (kg/m³)
bromine	3103
iodine	4933
astatine	

> Guided

(b)   Give a reason for your answer to (a).

> Look at the trend in density going down Group 7. This should help you to make a prediction for astatine.

Going down the Group, the density ....

....   **(1 mark)**

5   Fluorine, at the top of Group 7, exists as simple molecules. Each molecule contains two fluorine atoms.

(a)   Name the type of bond that exists between the atoms in a fluorine molecule, F_2.

> You have a choice of ionic bond, covalent bond or metallic bond.

....   **(1 mark)**

(b)   Fluorine has a low boiling point. Name the type of forces or bonds that are overcome when fluorine boils.

....   **(1 mark)**

Reactions of halogens

1 Hydrogen reacts with chlorine to produce hydrogen chloride gas, HCl(g).

(a) Balance the equation for this reaction.

$$H_2(g) + Cl_2(g) \rightarrow ......HCl(g)$$ **(1 mark)**

(b) What happens when hydrogen chloride gas is bubbled through water?

☐ **A** reacts vigorously, releasing oxygen   ☐ **C** dissolves to form an acidic solution

☐ **B** dissolves to form an alkaline solution   ☐ **D** dissolves to form a neutral salt
                                                        solution   **(1 mark)**

Guided

(c) Fluorine reacts with hydrogen in the cold and dark, but chlorine and hydrogen
must be exposed to sunlight in order to react. A mixture of hydrogen and
bromine reacts only if a flame is put in it.

Explain what this tells you about the | Look carefully at the conditions needed for
pattern of reactivity of the halogens. | hydrogen to react with the different halogens.

Going down Group 7, the elements become ....

I can tell this because the energy needed for them to start reacting ....

.... **(2 marks)**

2 Sodium burns in chlorine to produce sodium chloride. | Look at the information given to
                                                                      | deduce the reactants and products.
(a) Write the word equation for this reaction. | Do not mix words and formulae.

.... **(1 mark)**

(b) Bromine vapour reacts with hot iron wool. Red–brown iron(III) bromide, $FeBr_3$, is produced.
Predict the formula of iron(III) chloride, formed in the reaction between iron and chlorine.

.... **(1 mark)**

(c) Iodine vapour reacts slowly with hot iron wool to produce grey iron(II) iodide, FeI_2.

(i) Write the formula of the iron(II) ion and the | Make sure that you can recall the formulae
    formula of the iodide ion. | of elements, simple compounds and ions.

iron(II) ion .... **(1 mark)**

iodide ion .... **(1 mark)**

(ii) Write the balanced equation for the | Remember that iodine and the other halogens exist
     reaction between iron and iodine to | as diatomic molecules with the general formula X_2.
     form iron(II) iodide.

.... **(2 marks)**

3 Reactivity decreases going down Group 7, from fluorine to iodine. Explain, in terms of the
electronic configurations of their atoms, why fluorine is more reactive than chlorine.

Guided

Fluorine atoms are .... than chlorine atoms, so its outer shell is

....

and it gains an outer electron .... **(3 marks)**

Halogen displacement reactions

1   A student adds a few drops of aqueous bromine solution to a potassium iodide solution. Iodine and potassium bromide solution form.

> Answer C cannot be correct because distillation is a physical separation method, not a chemical reaction.

(a)   What type of reaction is this?

☐ **A** neutralisation   ☐ **B** precipitation   ☐ **C** distillation   ☐ **D** displacement

(1 mark)

(b)   Write a word equation for this reaction.

.... **(1 mark)**

2   A displacement reaction may happen when a halogen is added to a solution containing halide ions. The table shows results from an investigation with three halogens. A tick (✓) shows that displacement happens.

Halogen added	Halide ion in solution		
	Chloride	**Bromide**	**Iodide**
chlorine	not done	✓	✓
bromine	✗	not done	✓
iodine	✗	✗	not done

(a)   Use the results shown in the table to deduce the order of reactivity of these halogens.

The order of reactivity, starting with the most reactive, is ....

.... because chlorine displaces ....

.... but bromine displaces only ....

Iodine .... **(3 marks)**

(b)   Suggest a reason that explains why three possible experiments were not done in the investigation.

.... **(1 mark)**

(c)   Explain why iodine could displace astatine from sodium astatide solution.

> Astatine is beneath iodine in Group 7 of the periodic table.

....

.... **(2 marks)**

3   Fluorine is the most reactive halogen. It reacts with water to form hydrofluoric acid and oxygen.

(a)   Balance the equation for the reaction between fluorine and water.

......$F_2(g)$ + ...$H_2O(l) \rightarrow$ ......$HF(aq) + O_2(g)$ **(1 mark)**

(b)   Suggest a reason that explains why a mixture of fluorine and water cannot be used in displacement reactions.

> Look at the information given to you about fluorine and water.

.... **(1 mark)**

(c)   Fluorine gas is passed over filter paper soaked in sodium chloride solution. A displacement reaction occurs but it is difficult to detect chlorine forming. Suggest a reason that explains this observation.

> Think about what you know about the appearance of chlorine.

.... **(1 mark)**

The noble gases

1 Which of these properties explains why argon is used as a shield gas during welding?

   ☐ **A**   Argon is inert.   ☐ **C**   Argon has a low density.

   ☐ **B**   Argon is colourless.   ☐ **D**   Argon is a good conductor of electricity.   **(1 mark)**

Guided

2 Explain why helium is used as a lifting gas for party balloons and airships.

> There are two relevant properties. For each one, explain why it is important for this use of helium.

Balloons and airships rise because helium ....

Helium is inert, so .... **(2 marks)**

3 The table shows some information about the noble gases.

> Temperatures with less negative numbers are higher temperatures, so –10 °C is warmer than –20 °C.

Element	Melting point (°C)	Density (kg/m^3)
helium	–272	0.16
neon	–248	0.41
argon	–189	0.74
krypton	–157	1.47
xenon	–111	2.21
radon	–71	3.52

(a) Name the noble gas that has the lowest melting point.

.... **(1 mark)**

(b) Describe the trend in density in Group 0.

.... **(1 mark)**

(c) Oganesson, Og, was discovered early this century. It is placed in Group 0 of the periodic table, immediately below radon. Predict the melting point of oganesson, and explain your answer.

> What happens to the melting point going down Group 0?

....

.... **(2 marks)**

4 The table shows the electronic configurations of the first three noble gases.

Explain, in terms of their electronic configurations, why these noble gases are unreactive.

Element	Electronic configuration
helium, He	2
neon, Ne	2.8
argon, Ar	2.8.8

> Other than having two electrons in their innermost shells, what do these three elements have in common?

....

.... **(2 marks)**

Extended response – Groups

The table shows the first five elements in Groups 1 and 7 of the periodic table, in the order in which they are placed.

Group 1	Group 7
lithium	fluorine
sodium	chlorine
potassium	bromine
rubidium	iodine
caesium	astatine

Sodium chloride is an ionic compound, formed when sodium forms positively charged ions and chlorine forms negatively charged ions.

When heated, sodium reacts with chlorine to produce sodium chloride:

$$2Na(s) + Cl_2(g) \rightarrow 2NaCl(s)$$

Explain, in terms of electron transfer, how sodium and chlorine react together to form sodium chloride. You may, if you wish, include diagrams to show the electronic configurations of the atoms and ions involved. Use ideas about the trends in reactivity in Groups 1 and 7 to explain why the reaction between caesium and fluorine is very violent.

What are the trends in reactivity in Groups 1 and 7?

....

....

....

....

....

....

....

....

....

....

....

....

....

....

....   **(6 marks)**

This question also covers content from Topic 1 (Key concepts in chemistry). Remember that this topic is common to Paper 3 and Paper 4, and not just covered in Paper 3.

Rates of reaction

1 Reactions happen when reactant particles collide and the collisions have sufficient energy. Complete the table by placing a tick (✓) in each correct box to explain why the rate of reaction increases.

Change in reaction conditions	Frequency of collisions increased	Energy of collisions increased
increased concentration of a reacting solution		
increased pressure of a reacting gas		
increased temperature of reaction mixture		

(3 marks)

2 Marble is mostly calcium carbonate. It reacts with dilute hydrochloric acid to produce carbon dioxide. The graph shows the results of an investigation into the effect of changing the size of marble chips on the rate of reaction with $50\,cm^3$ of dilute hydrochloric acid. The same mass of chips, 3.2 g, was used each time.

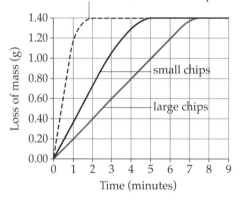

(a) State what happens to the surface area:volume ratio of a solid as the size of its particles is reduced.

.... (1 mark)

(b) Explain, in terms of the frequency of particle collisions, the expected result for powdered marble.

> Think about the surface area:volume ratio of the chips and the powder, and remember to look at the graph.

....

....

.... (2 marks)

3 (a) Describe the meaning of the term 'catalyst'.

> Guided

a substance that speeds up a reaction without altering the ....

and is unchanged ....

and is also unchanged .... (3 marks)

(b) Explain, in terms of activation energy, how a catalyst increases the rate of a reaction.

A catalyst provides an alternative ....

with a lower .... (2 marks)

(c) (i) State the name given to a biological catalyst. .... (1 mark)

   (ii) Give **one** example of a commercial use of a biological catalyst.

.... (1 mark)

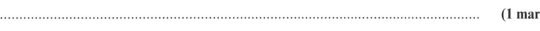

 **Investigating rates**

1 Sodium thiosulfate solution and dilute hydrochloric acid are clear, colourless solutions. They react together to form sodium chloride solution, water, sulfur dioxide and sulfur:

$$Na_2S_2O_3(aq) + 2HCl(aq) \rightarrow 2NaCl(aq) + H_2O(l) + SO_2(g) + S(s)$$

(a) Give **one** way in which the volume of a gas can be measured accurately.

Collect the gas in .... **(1 mark)**

(b) Sulfur dioxide is highly soluble in water. Explain why measuring the volume of sulfur dioxide collected is **not** an accurate way to determine the rate of this reaction.

> A *soluble* substance dissolves in a solvent such as water.

....

.... **(2 marks)**

(c) Suggest reasons that explain why the production of sodium chloride solution or water **cannot** easily be used to determine the rate of reaction.

> What changes, if any, would you expect to see when sodium chloride solution or water is produced?

....

.... **(2 marks)**

2 A student investigates how changes in the concentration affect the rate of the reaction between 50 cm³ of sodium thiosulfate solution and 5 cm³ of dilute hydrochloric acid. The diagram shows the method that she uses.

The student varies the concentration of sodium thiosulfate solution by diluting it with water, but she uses 70 g dm⁻³ of hydrochloric acid each time. The table shows her results.

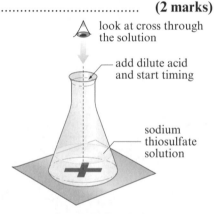

Concentration of $Na_2S_2O_3$(aq) added (g dm⁻³)	Time taken for cross on paper to disappear (s)	Relative rate of reaction, 1000/time (/s)
5	125	
15	42	24
25	25	

view through solution

As time goes on, the solution gets more cloudy. The cross 'disappears'.

(a) State **two** steps that the student takes to obtain valid results.

> Think about the variables, other than the concentration of sodium thiosulfate.

1 ....

2 .... **(2 marks)**

Guided (b) Complete the table to show the relative rate of each reaction.

> 🖩 **Maths skills** Use a calculator to calculate 1000 ÷ time for 5 g dm⁻³ and 25 g dm⁻³ of $Na_2S_2O_3$(aq). **(3 marks)**

(c) Describe what happens to the relative rate of reaction as the concentration of the sodium thiosulfate solution increases.

....

.... **(2 marks)**

Exam skills – Rates of reaction

1 Calcium carbonate reacts with dilute hydrochloric acid:

> calcium carbonate + hydrochloric acid →
> calcium chloride + water + carbon dioxide

A student adds lumps of calcium carbonate to an excess of dilute hydrochloric acid in a flask. He measures the change in mass that happens as carbon dioxide gas escapes from the flask. The table shows his results.

(a) Plot a graph of change in mass against time using the grid.   **(3 marks)**

Time (s)	Change in mass (g)
0	0.00
20	0.48
40	0.76
60	0.88
80	0.94
100	0.96
120	0.96

> Use × or + for each point, and draw a single line of best fit. The line does not have to be a straight line.

> Aim to plot each point to within half a square of its accurate position, or better.

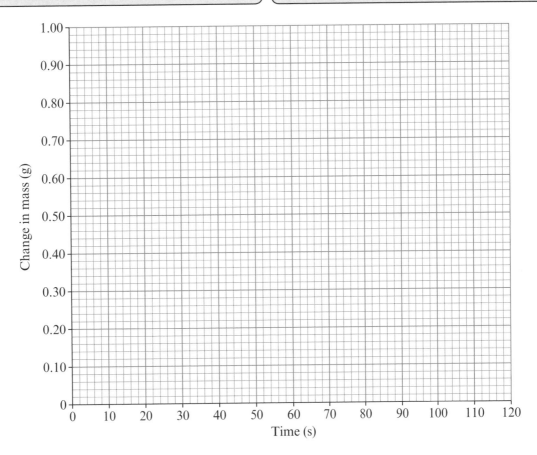

(b) Give the time taken for the reaction to finish. Explain your answer using information from the table or your graph.

time taken: ....

explanation: ....

.... **(2 marks)**

(c) The student repeats the experiment. He keeps all the conditions the same, but increases the temperature of the dilute hydrochloric acid. On the grid, sketch a line that the student should obtain for this experiment. Label this line **C**.

> You do not need to plot individual points for this line.

(2 marks)

Heat energy changes

Guided

1 Describe, in terms of energy transfers, the difference between an exothermic process and an endothermic process.

> Think about whether heat energy is taken in or given out in these processes.

In an exothermic change or reaction, heat energy is ....

but, in an endothermic change or reaction, heat energy is .... **(2 marks)**

2 Breaking bonds and making bonds involve energy transfers. Which row (**A**, **B**, **C** or **D**) in the table correctly describes these processes?

	Bond breaking	Bond making
☐ A	exothermic	exothermic
☐ B	exothermic	endothermic
☐ C	endothermic	exothermic
☐ D	endothermic	endothermic

> Answer A cannot be correct because one process is endothermic and the other is exothermic.

(1 mark)

3 Changes in heat energy occur when salts dissolve in water.

(a) Ammonium nitrate is dissolved in water. The temperature of the reaction mixture decreases.

State whether this process is exothermic or endothermic.

.... **(1 mark)**

(b) Give one type of reaction, which takes place in aqueous solution, that is always exothermic.

> Reactions that happen in solution include precipitation, neutralisation and displacement reactions.

.... **(1 mark)**

4 Magnesium reacts with dilute hydrochloric acid, forming magnesium chloride solution and hydrogen gas.

(a) Describe the measurements that you would take to confirm that the reaction is exothermic.

> **Practical skills** Outline what you would measure, the measuring apparatus and how you would use the results.

....

....

.... **(3 marks)**

(b) Balance the equation for this reaction, and include state symbols.

$Mg(......) + ...HCl(......) \rightarrow MgCl_2(......) + H_2(......)$ **(2 marks)**

Guided

(c) Explain, in terms of breaking bonds and making bonds, why this reaction is exothermic.

> Mention heat energy in your answer.

More ....

is released when bonds ....

than is needed to .... **(3 marks)**

Reaction profiles

1 Give the term used to describe the minimum energy needed to start a reaction.

.... **(1 mark)**

2 Methane burns completely in oxygen to form carbon dioxide and water vapour:

$$CH_4(g) + 2O_2(g) \rightarrow CO_2(g) + 2H_2O(g)$$

reactants   products

> **Guided**

The diagram shows a simple reaction profile for this reaction.

Explain, using information in the diagram, how you can tell that this reaction is exothermic.

> Is heat energy given out or taken in during this reaction?

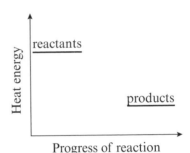

There is more stored energy in the ....

than in the ....

so, during the reaction, energy is .... **(2 marks)**

3 Carbon burns completely in excess oxygen to form carbon dioxide: $C(s) + O_2(g) \rightarrow CO_2(g)$

Complete the reaction profile for this reaction by showing the activation energy.

(2 marks)

> You will need to draw a curved line between the reactant and product lines, and add a labelled arrow.

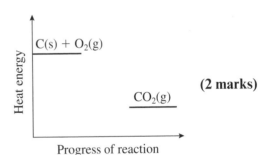

4 Calcium carbonate decomposes when heated strongly, forming calcium oxide and carbon dioxide:

$$CaCO_3(s) \rightarrow CaO(s) + CO_2(g)$$

The reaction is endothermic.

Label the reaction profile diagram to show the overall energy change and the activation energy.

> You will need to draw at least one dashed line and two labelled arrows.

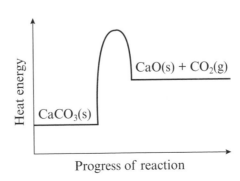

(2 marks)

Crude oil

1 (a) How long does it take for crude oil to form?

☐ **A** tens of years   ☐ **C** thousands of years

☐ **B** hundreds of years   ☐ **D** millions of years   **(1 mark)**

(b) Crude oil is described as mainly a complex mixture of:

☐ **A** hydrogen and carbon   ☐ **C** polymers

☐ **B** alkenes   ☐ **D** hydrocarbons   **(1 mark)**

(c) Crude oil is a **finite** resource. Explain what this means.

> Use the correct answer to (a) to help you.

....

.... **(1 mark)**

2 The diagrams show the structure of two compounds, hexane and cyclohexane.

hexane   cyclohexane

(a) The molecular formula of hexane is C_6H_{14}. Give the molecular formula of cyclohexane.

> Count the numbers of atoms of each element in the diagram of cyclohexane.

.... **(1 mark)**

Guided (b) Explain why hexane and cyclohexane are hydrocarbons.

They are compounds of ....

.... **(2 marks)**

3 Crude oil is an important source of fuels.

(a) Give **one** example of a fuel obtained from crude oil.

.... **(1 mark)**

(b) Petrochemicals are substances made from crude oil. They include polymers such as poly(ethene). Crude oil is a feedstock for the petrochemical industry. Explain the meaning of the term 'feedstock'.

....

....

.... **(2 marks)**

Fractional distillation

1   Crude oil is separated into simpler, more useful mixtures by fractional distillation. The diagram shows the main fractions obtained from crude oil.

crude oil →

→ gases
→ petrol
→ kerosene
→ diesel oil
→ fuel oil
→ bitumen

(a) Name the oil fraction that:

(i)   has the smallest number of carbon atoms and hydrogen atoms in its molecules

.... **(1 mark)**

(ii) contains substances with the highest boiling points

.... **(1 mark)**

(iii) is easiest to ignite

.... **(1 mark)**

(iv) has the highest viscosity.

> Viscosity is a measure of how difficult it is for a substance to flow.

.... **(1 mark)**

(b) Name the oil fraction that is used:

(i)   to surface roads and roofs

.... **(1 mark)**

(ii) as a fuel for aircraft.

.... **(1 mark)**

(c) Name two oil fractions that are used as fuels for cars.

.... **(2 marks)**

2   Most of the hydrocarbons in crude oil are members of a particular homologous series. Name this series.

.... **(1 mark)**

3   Describe how crude oil is separated using fractional distillation.

Guided

Oil is heated so that it ....

The vapours are passed into a column, which is hot at the ....

and cold at the .... Hydrocarbons rise, and ....

at different heights, depending on .... **(4 marks)**

145

Alkanes

1 Natural gas is a hydrocarbon fuel. It is mainly methane, CH_4. Which of the following substances **cannot** be produced when methane burns in air?

☐ **A** water   ☐ **B** carbon   ☐ **C** hydrogen   ☐ **D** carbon dioxide   **(1 mark)**

2 Which row correctly shows two features of a homologous series?

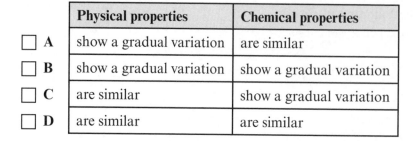

	Physical properties	Chemical properties
☐ **A**	show a gradual variation	are similar
☐ **B**	show a gradual variation	show a gradual variation
☐ **C**	are similar	show a gradual variation
☐ **D**	are similar	are similar

Physical properties include melting points and boiling points.

(1 mark)

3 The alkanes form a homologous series of hydrocarbons. The diagrams show the structures of the first two alkanes, methane and ethane.

$$H-\underset{\underset{H}{|}}{\overset{\overset{H}{|}}{C}}-H$$

methane

$$H-\underset{\underset{H}{|}}{\overset{\overset{H}{|}}{C}}-\underset{\underset{H}{|}}{\overset{\overset{H}{|}}{C}}-H$$

ethane

(a) Write the molecular formula of ethane.

The molecular formula of methane is shown in Question 1, and the structures of methane and ethane above.

.... **(1 mark)**

(b) The next alkane in the homologous series is propane, C_3H_8. Draw the structure of propane, showing all the covalent bonds.

Show each covalent bond as a straight line.

(1 mark)

> **Guided**

(c) Deduce how the molecular formula of an alkane differs from its neighbouring compounds.

Compare the three molecular formulae. It may also help to compare the three structures.

Going from one alkane to the next, the molecular formula changes by ....

.... **(1 mark)**

4 (a) State the general formula for the alkanes.

What pattern is shown by the molecular formulae for methane in question 1, and for ethane and propane in Question 3?

> **Guided**

C_nH .... **(1 mark)**

(b) Predict the molecular formula for hexane, an alkane that has six carbon atoms in its molecules.

.... **(1 mark)**

Incomplete combustion

1 Petrol is a hydrocarbon fuel. When it burns in air, waste products form and energy is transferred to the surroundings.

(a) Which releases the most energy from 1 dm³ of petrol: complete combustion or incomplete combustion?

.... **(1 mark)**

Guided

(b) Complete the table by placing a tick (✓) in each correct box to show the products formed in each type of combustion.

	Incomplete combustion	Complete combustion
Water		✓
Carbon		
Carbon monoxide		
Carbon dioxide	✓	

(2 marks)

2 Diagrams **A** and **B** show two flames produced by a Bunsen burner. The air hole is closed in diagram **A** but open in diagram **B**. If the air hole is closed, less oxygen can enter the chimney to mix with the natural gas.

Explain which flame (**A** or **B**) will coat the bottom of a gauze mat with a black substance.

....

....

A   B

.... **(2 marks)**

3 The combustion of diesel oil can produce carbon particles and carbon monoxide gas.

(a) Give a reason that explains why carbon particles may be harmful to health if breathed in.

.... **(1 mark)**

(b) Balance the equation below, which models a combustion reaction of benzene (a liquid hydrocarbon).

> First balance the number of water molecules needed, then the number of oxygen molecules needed.

$C_6H_6 + ...O_2 \rightarrow 2C + 3CO + CO_2 + ...H_2O$ **(1 mark)**

(c) Explain why carbon monoxide is toxic.

When breathed in, carbon monoxide combines with ....

so .... **(2 marks)**

Acid rain

1 (a) Complete the diagram below using labels from the box.

| acid rain | power station | rain cloud | acidic gases | distant city |

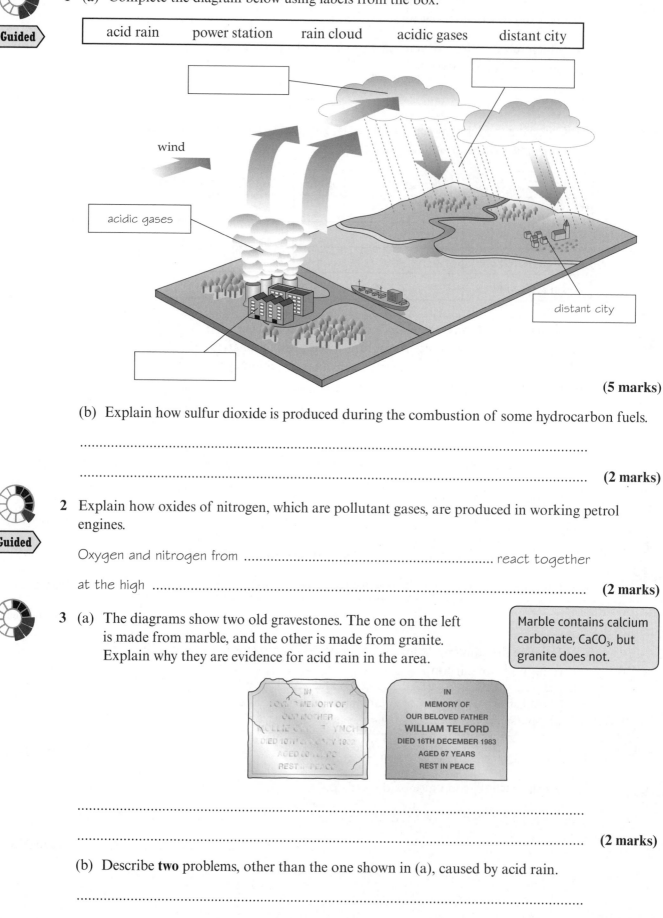

(5 marks)

(b) Explain how sulfur dioxide is produced during the combustion of some hydrocarbon fuels.

....

.... **(2 marks)**

2 Explain how oxides of nitrogen, which are pollutant gases, are produced in working petrol engines.

Oxygen and nitrogen from .... react together

at the high .... **(2 marks)**

3 (a) The diagrams show two old gravestones. The one on the left is made from marble, and the other is made from granite. Explain why they are evidence for acid rain in the area.

> Marble contains calcium carbonate, $CaCO_3$, but granite does not.

....

.... **(2 marks)**

(b) Describe **two** problems, other than the one shown in (a), caused by acid rain.

....

.... **(2 marks)**

Choosing fuels

1 Petrol, kerosene and diesel oil are fossil fuels.

(a) State the name of the substance from which these fuels are obtained.

.... **(1 mark)**

▷ Guided ▷ (b) Give **one** example of how each fuel is used.

Petrol is used as a fuel for cars. Kerosene is used as a ....

and diesel oil is used .... **(3 marks)**

2 Crude oil and natural gas are finite resources because they take a very long time to form, or are no longer being made. Methane is a non-renewable fossil fuel that is found in natural gas.

> Non-renewable and finite have different meanings. Do not answer by writing 'it is not renewable'.

State why methane is described as **non-renewable**.

.... **(1 mark)**

3 Hydrogen and petrol may both be used as fuels for cars.

(a) Complete the balanced equation for the reaction between hydrogen and oxygen.

> There is only one product.

$2H_2 + O_2 \rightarrow$ .... **(1 mark)**

(b) Petrol is a complex mixture of hydrocarbons. Name one product of the **complete combustion** of petrol that is **not** produced when hydrogen burns.

.... **(1 mark)**

4 The table shows some information about hydrogen gas and liquid petrol.

Fuel	Energy released by 1 dm³ of fuel (MJ)	Energy released by 1 kg of fuel (MJ)
hydrogen	0.012	142
petrol	35	47

(a) Using information from the table, identify an advantage of using:

(i) petrol rather than hydrogen as a fuel for cars

.... **(1 mark)**

(ii) hydrogen rather than petrol as a fuel for cars.

.... **(1 mark)**

(b) Suggest a reason that explains why there is an advantage to storing hydrogen in car fuel tanks as a liquid, rather than as a gas.

> Think about the closeness of particles in liquids and gases.

....

.... **(1 mark)**

Cracking

1 In the diagram below, a cracking reaction is modelled using the structures of the molecules.

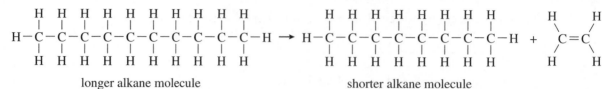

longer alkane molecule   shorter alkane molecule

(a) Name the **type** of hydrocarbon shown by the smallest molecule above.

.... **(1 mark)**

> **Guided**

(b) Write the balanced equation, using molecular formulae, for this cracking reaction.

> Count the carbon atoms and hydrogen atoms in each molecule to work out the formulae needed.

$C_{10}H$ .... **(2 marks)**

2 Alkanes and alkenes form two different homologous series. Which row in the table correctly describes each type of hydrocarbon?

	Alkanes	Alkenes
☐ A	saturated	contain only C–H and C–C bonds
☐ B	unsaturated	contain only C–H and C–C bonds
☐ C	contain only C–H and C–C bonds	saturated
☐ D	contain only C–H and C–C bonds	unsaturated

(1 mark)

3 The apparatus shown in the diagram can be used to crack the alkanes in liquid paraffin.

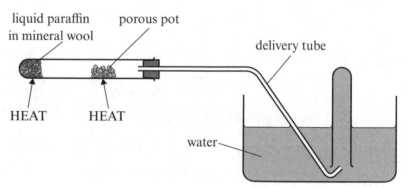

liquid paraffin in mineral wool   porous pot

delivery tube

HEAT   HEAT

water

(a) State the purpose of the porous pot.

.... **(1 mark)**

> **Guided**

(b) Explain what is meant by **cracking**.

a reaction in which larger alkanes are broken down into ....

....

.... **(2 marks)**

4 Crude oil is separated into more useful mixtures called fractions by fractional distillation. Give **two** reasons why an oil refinery may crack the fractions containing larger alkanes.

> **Guided**

Smaller hydrocarbons are ....

Cracking helps to match .... **(2 marks)**

Extended response – Fuels

Camping gas is a mixture of propane and butane, obtained from crude oil. It is a rainy day and some campers are making tea inside their tent. Incomplete combustion of the camping gas could occur if the campers do not take adequate precautions.

> How does incomplete combustion occur?

Explain how incomplete combustion of hydrocarbons such as propane and butane occurs, and the problems that it can cause in a situation similar to this one. You may include a balanced equation in your answer.

> Which is a more efficient use of fuels, complete or incomplete combustion, and why does this matter?

> What are the products of the incomplete combustion of hydrocarbons? What problems do they cause?

....
....
....
....
....
....
....
....
....
....
....
....
....
....
....
....
....   **(6 marks)**

> Carbon dioxide may be produced during incomplete combustion as well as during complete combustion, and so does not explain the problems that this gas causes.

The early atmosphere

1   The gases that formed the Earth's earliest atmosphere are thought to have come from:

☐ **A** combustion

☐ **B** volcanic activity

☐ **C** photosynthesis

☐ **D** condensation   **(1 mark)**

2   The pie chart shows the possible percentages of three gases in the Earth's early atmosphere.

Key
■ carbon dioxide
■ water vapour
☐ other gases

10%
80%

📊 **Maths skills** Use the key to identify the sector that represents carbon dioxide.

(a)   State the percentage of carbon dioxide in the early atmosphere, as shown by this pie chart.

....   **(1 mark)**

(b)   Explain how oceans formed as the early Earth cooled.

....

....   **(2 marks)**

(c)   The Earth's atmosphere today contains about 0.04% carbon dioxide. Explain how the oceans contributed to the decrease in the percentage of carbon dioxide in the atmosphere.

Think about whether carbon dioxide is soluble or insoluble in water.

....

....   **(2 marks)**

3   The Earth's atmosphere today contains more oxygen than its early atmosphere did.

Write down what you would do and what you would observe.

(a)   Describe the chemical test for oxygen.

....

....   **(2 marks)**

▷ Guided ▷   (b)   Explain why the percentage of oxygen in the atmosphere has gradually increased.

In your answer, include the name of the process involved.

The growth of primitive plants used ....

and released ....

by the process of ....   **(3 marks)**

Greenhouse effect

1 Which of these gases are both greenhouse gases?

☐ **A** oxygen and carbon dioxide

☐ **B** carbon dioxide and water vapour

☐ **C** nitrogen and carbon dioxide

☐ **D** oxygen and nitrogen   **(1 mark)**

Guided

2 The table shows processes involved in the greenhouse effect. Add a number to each box to order the processes from **1** (first) to **4** (last).

Process	Order (1–4)
Gases in the atmosphere absorb heat radiated from the Earth.	
Released energy keeps the Earth warm.	4
Heat is radiated from the Earth's surface.	
Gases in the atmosphere release energy in random directions.	

(3 marks)

3 Carbon dioxide is described as a greenhouse gas.

(a) Name another greenhouse gas, often released as a result of livestock farming.

.... **(1 mark)**

(b) Explain why the use of fossil fuels causes the release of carbon dioxide.

> What type of substance is found in fossil fuels such as petrol? What happens when they are burned?

....

.... **(2 marks)**

4 The graphs show how the mean global temperature and the percentage of carbon dioxide in the atmosphere have changed over the last 220 000 years.

— difference in temperature

▬ percentage of CO_2 in the air

200 1850 100   50   0
Thousands of years before today

(a) State what happens, in general, to the mean global temperature as the percentage of carbon dioxide in the air increases.

.... **(1 mark)**

Guided

(b) State what is meant by **global warming**.

> Do not use the words **global** or **warming** in your answer.

a worldwide .... **(1 mark)**

(c) State **two** environmental effects of global warming.

1 .... **(1 mark)**

2 .... **(1 mark)**

Extended response – Atmospheric science

The graphs show the change in mean global temperature, and the concentration of carbon dioxide in the atmosphere, between the years 1850 and 2005.

A higher temperature than the mean temperature gives a positive temperature change on the graph.

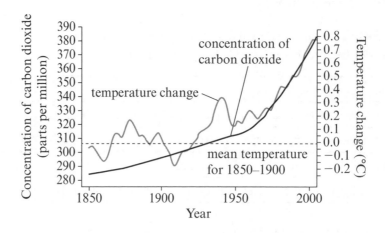

Evaluate whether these graphs provide evidence that human activity is causing the Earth's temperature to increase. In your answer, explain how carbon dioxide acts as a greenhouse gas, and describe processes that release carbon dioxide or remove it from the atmosphere.

....

....

....

....

....

....

....

....

....

....

....

....

.... **(6 marks)**

You should be able to evaluate evidence for human activity causing climate change. This could include correlations between the change in atmospheric carbon dioxide concentration, the consumption of fossil fuels and temperature change.

Key concepts

1 Complete the table for units of physical quantities and their abbreviations.

ampere		joule		pascal	Pa		
		watt		newton		ohm	

(2 marks)

2 Explain the difference between a base unit and a derived unit.

> Use the words 'independent' and 'made up' in your answer.

....

.... **(2 marks)**

3 Convert each quantity.

(a) 750 grams to kilograms

> **Maths skills** 1000 g = 1 kg; 1000 W = 1 kW; 60 s = 1 minute;
> 1000 mm = 1 m; 1 000 000 J = 1 MJ

.... kg   **(1 mark)**

(b) 0.75 kilowatts to watts

.... W   **(1 mark)**

(c) 25 minutes to seconds

.... s   **(1 mark)**

4 A frequency of 2.5 kHz is 2500 Hz or, in standard form, 2.5×10^3 Hz.

(a) Write 8 nm in metres, m.

.... **(1 mark)**

(b) Write 8 nm in m in standard form.

.... **(1 mark)**

5 Calculate the speed of a car that takes 10.5 s to travel 75 m. Give your answer to 5 significant figures.

> **Maths skills** Look at the number that follows the significant figure you are asked to consider
> (in this case the 5th one). If it is greater than 5, round the 5th figure up; if it is
> less then round down, e.g. 1.23076923 would become 1.2308 to 5 significant figures.

$s = d \div t$ so .... m ÷ .... s = ... m/s

so to 5 significant figures

speed = ... m/s   **(2 marks)**

Scalars and vectors

1 (a) Identify the correct category for each quantity listed below. Write your answers in the table.

acceleration   displacement   speed
temperature   mass   force   velocity
distance

> A scalar has only a magnitude (size) but a vector has both a magnitude **and** a direction.

Scalars	Vectors

(2 marks)

> **Guided**

(b) (i) Give **one** example of a scalar from your table and explain why it is a scalar.

.... is a scalar because ....

.... **(2 marks)**

(ii) Give **one** example of a vector from your table and explain why it is a vector.

.... is a vector because ....

and .... **(2 marks)**

2 At the swimming pool, two swimmers are practising for a swimming gala. They swim from opposite ends of the pool. The first swimmer dives in from the left side and swims the length of the pool at a velocity of 1.3 m/s. The second swimmer then swims from the right at a velocity of −1.4 m/s.

(a) Explain why velocity is used in this example instead of speed.

....

.... **(2 marks)**

(b) Explain why the second swimmer's velocity has a negative value.

.... **(1 mark)**

3 (a) Which of the following is **not** a scalar?

☐ **A** energy   ☐ **C** mass

☐ **B** temperature   ☐ **D** weight   **(1 mark)**

(b) Give a reason for your answer to (a).

....

.... **(1 mark)**

4 (a) Write the letter of each runner shown in the picture next to a suitable velocity in the table below.

−3 m/s	
5 m/s	
4.5 m/s	

(2 marks)

(b) Give a reason for your choice.

.... **(1 mark)**

Speed, distance and time

1 The distance/time graph shows a runner's journey from his home to the park.

(a) State the letter that corresponds to the part of the runner's journey where he:

   (i) stops   ....   **(1 mark)**

   (ii) runs fastest.   ....   **(1 mark)**

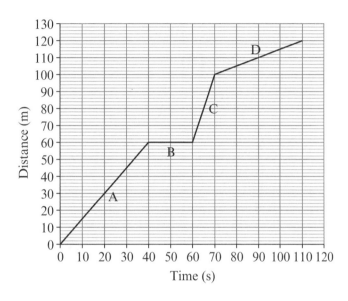

Guided

(b) Calculate the runner's speed in part A of his journey.

In part A, he travels .... m in .... s.

speed = distance ÷ ....

speed = ... m/s   **(3 marks)**

(c) Explain how the graph shows a faster speed compared with a slower speed in different parts of the runner's journey.

....

.... **(2 marks)**

2 The lift in a wind turbine tower takes 24 s to go from the ground to the generator 84 m above.

> Speed = distance ÷ time only. Velocity is speed in a given direction.

(a) Calculate the average speed of the lift. State the unit.

speed = ... unit .... **(3 marks)**

(b) State the average velocity of the lift.

.... **(1 mark)**

3 A jogger runs through a park at a constant speed of 5 m/s covering a distance of 400 m.

Calculate the time it takes for the jogger to run the 400 m.

time = ... s **(2 marks)**

Equations of motion

1 Identify the unit for acceleration.

☐ **A** ms^2

☐ **C** m/s^2

☐ **B** m^2

☐ **D** m/s

2 Draw a line from each symbol to its correct description. One has been done for you.

> Initial velocity occurs **before** acceleration causes a change in velocity.

Symbol
v
u
a
x
t

Description
acceleration
time
distance
final velocity
initial velocity

(2 marks)

3 (a) A racing car takes 8 seconds to speed up from 15 m/s to 25 m/s. Calculate its acceleration.

> You may find the equation $a = v - u \div t$ useful. Remember that initial velocity $= u$ and final velocity $= v$.

acceleration = ... m/s^2 **(3 marks)**

Guided

(b) The racing car now accelerates at the same rate for 12 seconds, from 25 m/s to a higher velocity. It travels 300 m during this time. Calculate its final velocity.

$v^2 = u^2 + 2 \times a \times x$

$= ....^2 + 2.............. \times ....$

$v^2 =$

so $v =$

velocity = ... m/s **(3 marks)**

Velocity/time graphs

1   The velocity/time graph shows how the velocity of a car changes with time.

    (a)   This graph can be used to analyse the car's journey. Choose **one** correct statement that describes what else the graph shows as well as acceleration.

       ☐   **A**   the distance the car travelled

       ☐   **B**   how long the car was stopped

       ☐   **C**   where the car travelled

       ☐   **D**   the constant velocity of the car

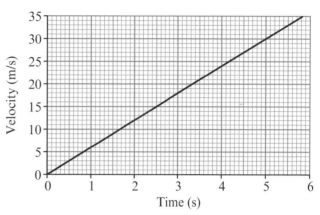

                                                                     (1 mark)

    (b)   Draw a triangle on the graph to show the acceleration and the time taken.   **(1 mark)**

> **Guided**

    (c)   Calculate the acceleration of the car as shown in the graph.

       change in velocity = ... m/s, time taken for the change = ... s

$$\text{acceleration} = \frac{\text{change in velocity}}{\text{time taken}} =$$

                             acceleration = ... m/s²   **(2 marks)**

    (d)   Use the graph to calculate the distance travelled by the car in the first 5 s.

                                  ┌──────────────────────────────┐
                                  │ Work out the area under the graph. │
                                  └──────────────────────────────┘

                                      distance = ... m   **(2 marks)**

2   A cyclist takes 5 seconds to reach maximum velocity of 4 m/s, from an initial velocity of 0 m/s, moving in a straight line.

> **Guided**

    (a)   Calculate the cyclist's acceleration.

       change in velocity =

       so

                            acceleration = ... unit ....   **(3 marks)**

    (b)   The cyclist travels at constant velocity for 15 seconds and then takes another 15 seconds to slow down to a stop. Sketch a graph of the journey and use this to explain how the total distance travelled could be calculated.

      ....

      ....

      ....

      ....   **(3 marks)**

Determining speed

1   Draw a line from each activity to its correct speed. One has been done for you.

Activity	Speed
commuter train	330 m/s
running	1.5 m/s
speed of sound in air	3.0 m/s
walking	55 m/s

(1 mark)

Guided

2   The diagram shows a light gate being used to measure the speed of a model vehicle.

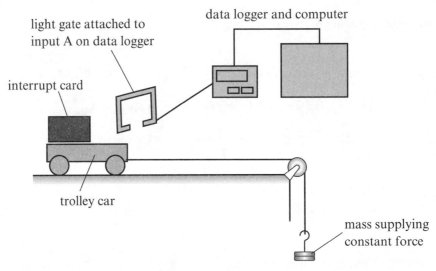

light gate attached to
input A on data logger

data logger and computer

interrupt card

trolley car

mass supplying
constant force

(a) Describe how the interrupt card can be used to measure the speed of the trolley car.

The light beam is ....

as it enters the light gate and this starts the timer. When the card has passed

through ....

and ....   **(3 marks)**

(b) State how the speed is found using this method.

....

....   **(1 mark)**

3   Suggest why it is better to use light gates
and a computer than to use a stopwatch
and a ruler to measure the speed of a toy car.

> Light gates can measure instantly so
> computers can calculate speeds over very
> short distances. Consider this advantage
> over measurement by a person.

....

....

....

....   **(3 marks)**

Newton's first law

1　Explain what is meant by a resultant force.

....

.... **(2 marks)**

2　A speed skater is standing on the ice waiting for the start of a race.

> Guided

(a)　Describe the action and reaction forces acting on the skater and her skates.

The action is the .... and the

reaction is .... **(2 marks)**

(b)　The race begins and the skater pushes against the ice producing a forward thrust on the skates of 30 N. There is resistance from the air of 10 N and friction on the blades of 1 N. Calculate the resultant force.

resultant force = positive direction – negative direction so

force = ... N　**(2 marks)**

(c)　During the race the resistive forces become equal to the forward thrust. Describe what happens to the velocity of the skater.

....

.... **(2 marks)**

(d)　At the end of the race the skater stops skating. Describe what happens next before the skater comes to a halt.

....

.... **(2 marks)**

3　A space probe falls towards the Moon. In the Moon's gravitational field the probe has a weight of 1700 N. The probe fires rockets giving an upward thrust of 1900 N.

(a)　Calculate the resultant force on the space probe.

resultant force = ... N　**(2 marks)**

(b)　Explain the changes in the probe's velocity.

....

....

.... **(2 marks)**

Newton's second law

1 In an experiment a student pulls a force meter attached to a trolley along a bench. The trolley has frictionless wheels. The force meter gives a reading of 5 N.

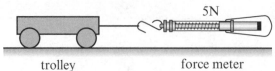

trolley   force meter

(a) Describe what happens to the trolley.

The trolley will ....

in the direction .... **(2 marks)**

(b) The student stacks some masses on the trolley and again pulls it with a force of 5 N. Explain why the trolley takes longer to travel the length of the bench.

The acceleration is .... because...

.... **(2 marks)**

2 When the Soyuz spacecraft travels to the International Space Station, it is accelerated by rockets. The spacecraft has a mass of 3000 kg and the peak acceleration of the spacecraft is 39 m/s^2.

(a) Calculate the peak resultant force acting on the spacecraft. State the unit.

$F = m \times a$

force = ... unit .... **(3 marks)**

(b) State the direction in which the force acts.

.... **(1 mark)**

3 A Formula One racing car has a mass of 640 kg. A resultant force of 10 500 N acts on the car.

(a) Calculate the acceleration on the racing car. State the unit.

acceleration = ... unit .... **(3 marks)**

(b) Explain what will happen to the acceleration of the car as its fuel tank empties, assuming the resultant force in forward direction remains constant.

....

.... **(2 marks)**

Weight and mass

1   Which of the following is a description of weight?

   ☐   **A**   Weight is a type of force.

   ☐   **B**   Weight is measured in kilograms (kg).

   ☐   **C**   Weight is a measure of mass.

   ☐   **D**   Weight is calculated by mass ÷ gravitational field strength.   **(1 mark)**

2   The lunar roving vehicle (LRV), driven by astronauts on the Moon, has a mass of 210 kg on Earth. State the mass of the unchanged LRV on the Moon. Give a reason for your answer.

   The mass of the LRV on the Moon is .... kg

   because ....

   ....

   ....   **(2 marks)**

3   (a)   Calculate the total weight of a backpack of mass 1 kg, containing books with a mass of 2 kg and trainers with a mass of 1.5 kg. Take gravitational field strength (*g*) to be 10 N/kg.

   > Use the equation relating weight to mass and gravitational field strength.

   weight = ... N   **(3 marks)**

   (b)   The book and trainers are removed from the bag and replaced with a sports kit. The weight of the bag is now 30 N. Calculate the mass of the sports kit.

   > You will need to rearrange the equation used in (a).

   mass = ... kg   **(2 marks)**

Force and acceleration

A ramp, a trolley, masses and electronic light gates can be used to investigate the interrelationship of force, mass and acceleration.

1 State **one** advantage of using electronic measuring equipment to determine acceleration compared with using a ruler and stopwatch.

....

....

....

.... **(2 marks)**

2 Describe how acceleration changes with mass, for the same force.

.... **(1 mark)**

3 Explain why it is necessary to use two light gates when measuring acceleration in this experiment.

> Guided

Acceleration is calculated by the change in speed ÷ time taken, so ....

....

....

.... **(2 marks)**

4 Describe the conclusion that can be drawn from this experiment.

> Guided

For a constant slope, ...

....

....

.... **(2 marks)**

5 Suggest **one** hazard associated with this experiment and **two** safety precautions that could be taken to minimise the risk of harm to the scientist.

> Consider the potential dangers of using accelerated masses or electrical equipment.

....

....

....

....

....

....

.... **(3 marks)**

Newton's third law

1 Select the statement that summarises Newton's third law.

☐ **A** For every action there is a constant reaction.

☐ **B** The action and reaction forces are different due to friction.

☐ **C** Reaction forces may be stationary or at constant speed.

☐ **D** For every action there is an equal and opposite reaction.   **(1 mark)**

2 Rockets are used to carry people into space to the International Space Station. Explain how Newton's third law can be used to describe the motion of the rocket.

> The rocket pushes out hot gases that exert a force.

....

....

....

....

....

....   **(3 marks)**

3 (a) A runner stretches her calf muscle, as shown in the diagram, against a brick wall. Explain this using Newton's third law.

....

....

....

....   **(2 marks)**

Guided

(b) If the brick wall is replaced with a fabric curtain explain, using Newton's third law, why the athlete would fall forwards if she pushed with the same force as she used on the brick wall.

The curtain is not rigid and so would not ....

....

....

....   **(2 marks)**

Human reaction time

1 (a) Reaction time is an important consideration in driving a vehicle safely. Which is the distance travelled due to the reaction time of a driver?

   ☐ **A** overall stopping distance

   ☐ **B** thinking distance

   ☐ **C** braking distance

   ☐ **D** reaction distance   **(1 mark)**

   (b) Suggest a factor that may influence the reaction time of a driver.

   .... **(1 mark)**

2 Explain how human reaction time is related to the brain.

Guided

Human reaction time is the ....

....

It is related to how quickly ....

.... **(2 marks)**

3 Explain how to measure human reaction time using a ruler.

   | Outline the main points in the 'drop and grab' test. |

   ....

   ....

   ....

   ....

   ....

   .... **(3 marks)**

4 Name a profession that relies on fast reaction times and give a reason why this is important.

   ....

   ....

   ....

   .... **(2 marks)**

Stopping distance

1 Write the word equation used to calculate overall stopping distance.

.... **(1 mark)**

2 Complete the table below to summarise the factors
that affect overall stopping distance.

> Separate the factors that
> may affect the reaction
> time of a driver from those
> that affect the vehicle.

Guided

> Thinking distance is affected by the driver; braking
> distance is affected by the car or conditions.

Factors increasing overall stopping distance	
Thinking distance will increase if	**Braking distance will increase if**
the car's speed increases	the car's speed increases
the driver is distracted	the road is icy or wet

(2 marks)

3 Explain why the overall stopping distance of a car with worn tyres is different from a car with new
tyres when driving in wet weather.

....

....

....

.... **(2 marks)**

4 Dom drives to the airport alone to collect his family and their luggage. Complete the following
paragraph to describe factors affecting the stopping distance of his vehicle.

When Dom drives alone, the total .... of the vehicle will be

.... than when his family and their luggage are in the car. For the same

.... , Dom will need to allow a ....

with his family in the car than when he was travelling alone. **(4 marks)**

Had a go ☐   Nearly there ☐   Nailed it! ☐

Extended response – Motion and forces

A skydiver leaves the training aeroplane and accelerates before opening the parachute. After opening the parachute the speed of the skydiver reduces significantly so that he lands safely. Explain the forces acting on the skydiver before and after opening the parachute and suggest a reason for the skydiver being in the position shown in the photograph.

You should try to use the information given in the question.

You will be more successful in extended response questions if you plan your answer before you start writing. The question asks you to give a detailed explanation of the forces acting on the skydiver from when he leaves the plane to when he lands on the ground. Think about:

- the forces acting on the skydiver causing him to accelerate
- how the forces change during the descent
- the effect of air resistance
- the effect of change in surface area
- the effect of balanced forces
- how the skydiver is able to land at a safe velocity.

You should try to use the information given in the question.

....

....

....

....

....

....

....

....

....

.... **(6 marks)**

Energy stores and transfers

1 Which of the following is an energy store?

☐ **A** electrical

☐ **B** light

☐ **C** radiation

☐ **D** thermal   **(1 mark)**

2 Explain how energy transfers can be represented using a diagram.

An energy transfer diagram shows both the ....

....

....

....

....   **(3 marks)**

3 A footballer has a breakfast of cereal and toast before setting off for a training session at the club. Complete the flow chart to show how energy is transferred to other stores.

   [Write the correct store(s) of energy in each space.]

....
energy in the breakfast

↓

....
energy of the footballer

↓

....
energy dissipated to the
surroundings   **(3 marks)**

4 The bar graphs below illustrate energy stores before each energy transfer occurs. Add bars to the second graph to show changes in the energy stores after the energy transfer has occurred. The graphs represent the energy of a bobsleigh at the top of a slope and halfway down the slope.

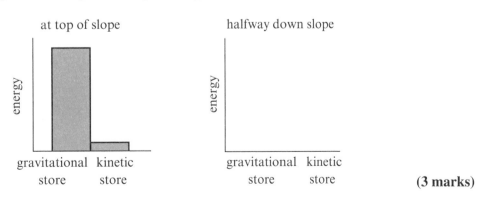

Efficient heat transfer

1 Identify the most suitable material, from the table below, for building an energy-efficient garage. Give a reason for your answer.

> The larger the relative thermal conductivity, the more heat will be conducted through the material.

Material	Relative thermal conductivity
brick	1.06
concrete	1.00
sandstone	2.20
granite	2.75

....

....

.... **(2 marks)**

2 Some houses are built with very thick walls. Explain how these walls help to keep the houses warm in the winter.

....

....

....

.... **(2 marks)**

3 A crane lifts a box to the top of a building. 1 000 000 joules is transferred to the gravitational store when the box is moved from the bottom to the top of the building. The crane uses fuel with 4 000 000 joules in a chemical store. Calculate the efficiency of the crane.

> **Guided**

$$\text{efficiency} = \frac{\text{useful energy transferred by the crane}}{\text{total energy supplied to the crane}}$$

useful energy transferred =

total energy used by the crane =

Efficiency = ... **(2 marks)**

4 (a) The motor in a food blender has an efficiency of 20%. The motor transfers 40 joules per second into the kinetic store. Calculate the energy that is transferred to the motor each second.

energy transferred each second = ... J **(3 marks)**

(b) State the power of the motor. Give the unit.

power = ... unit .... **(1 mark)**

Energy resources

1   Name **three non-renewable** energy resources.

.... **(1 mark)**

2   Identify the **renewable** energy resources from their descriptions given below.

(a) generates electricity from water trapped by a dam and then flows down a pipe

.... **(1 mark)**

(b) uses the rise and fall of the tide to generate electricity

.... **(1 mark)**

(c) uses the wind to generate electricity

.... **(1 mark)**

3   Some of the sources of renewable energy listed below are only available at certain times, whereas other sources can be used at any time.

Guided

| hydroelectric | tidal | solar | wind | geothermal |

(a) State the sources of renewable energy in the list that are always available.

hydroelectric and .... **(1 mark)**

(b) Explain why it is an advantage to have energy sources available at any time.

> Think about how the weather affects some renewable energy sources.

Demand is greatest at certain ....

Demand may be high when ....

.... **(2 marks)**

4   Compare the way geothermal and oil-fired power stations are used to generate electricity.

....

....

.... **(3 marks)**

5   A hydroelectric power station is used to produce electricity when demand is high.

(a) Explain why the hydroelectric power station is a reliable producer of electricity.

....

.... **(2 marks)**

(b) Give **one** reason why we cannot use hydroelectric power stations in more places in the UK.

....

.... **(1 mark)**

Patterns of energy use

> **Guided**

1 The graphs show patterns of energy use and human population growth.

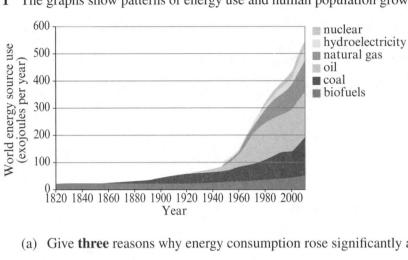

(a) Give **three** reasons why energy consumption rose significantly after the year 1900.

After 1900 the world's ....

There was development in ....

and .... **(3 marks)**

(b) (i) Identify which category of energy resources has been the main contributor to world energy consumption since the year 1900.

.... **(1 mark)**

(ii) Suggest **two** reasons why the consumption of energy resources has increased in the developed world.

....

.... **(2 marks)**

(iii) Suggest a reason why nuclear energy resources did not appear before the 1900s.

.... **(1 mark)**

(iv) Identify a renewable resource from the graph that makes use of a change in gravitational potential energy.

.... **(1 mark)**

2 If the patterns in energy consumption are similar to the patterns in the world's population growth, discuss the issues resulting from the continuing use of energy in the way shown in the graph in Q1.

> Consider finite non-renewable resources and increasing demand due to population, transport and industrial growth.

....

....

....

....

.... **(4 marks)**

Potential and kinetic energy

1 (a) State how an object gains gravitational potential energy.

.... **(1 mark)**

(b) State the **two** variables that can cause a change in the amount of gravitational potential energy of an object.

.... **(2 marks)**

2 Identify the correct equation for calculating gravitational potential energy change.

☐ **A** GPE = $m \times v \times h$ ☐ **C** GPE = $m \times F \times a$

☐ **B** GPE = $\frac{1}{2} m \times v^2$ ☐ **D** ΔGPE = $m \times g \times \Delta h$ **(1 mark)**

3 Calculate the change in gravitational potential energy of a drone that has a mass of 2 kg and is raised by 25 m. State the unit.

gravitational potential energy = ... unit .... **(3 marks)**

4 Add the correct labels, A, B and C, to the diagram to show where the maximum and minimum gravitational potential energy (GPE) and kinetic energy (KE) occur in a swinging pendulum.

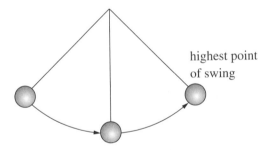

highest point
of swing

A	maximum KE	minimum GPE
B	no KE	maximum GPE
C	maximum GPE	no KE

(2 marks)

5 Calculate the kinetic energy of a cyclist and her bicycle, with combined mass of 70 kg, travelling at 6 m/s.

> You may find this equation useful:
>
> KE = $\frac{1}{2} \times m \times v^2$

kinetic energy = ... J **(2 marks)**

Extended response – Conservation of energy

Millie plays on a swing in the park. The swing seat is initially pulled back by her friend to 30° to the vertical position and then released. Describe the energy changes in the motion of the swing.

Your answer should also explain, in terms of energy, why the swing eventually stops.

You will be more successful in extended response questions if you plan your answer before you start writing.

The question asks you to give a detailed explanation of the energy changes as the swing moves backwards and forwards. Think about:

- how gravitational potential energy changes as the swing is pulled back
- points at which gravitational potential energy (GPE) and kinetic energy (KE) are at maximum and at 0
- where some energy may be lost from the system
- why the swing will eventually stop.

You should try to use the information given in the question.

....

....

....

....

....

....

....

....

....

....

....

....

....

....

.... (6 marks)

Waves

1 The table below lists statements about transverse and longitudinal waves. Identify which type of waves are described by writing T (transverse), L (longitudinal) or B (both) next to each statement.

> Consider whether particles are involved and how they move.

Sound waves are this type of wave.		They have amplitude, wavelength and frequency.	
All electromagnetic waves are this type of wave.		Seismic P waves are this type of wave.	
Particles oscillate in the same direction as the wave.		They transfer energy.	

(3 marks)

2 The diagram below shows a wave travelling through a medium.

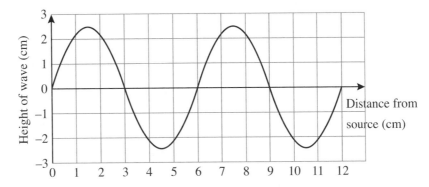

(a) What is the amplitude of the wave in the diagram?

☐ **A** 0.05 m   ☐ **B** 0.025 m   ☐ **C** 0.12 m   ☐ **D** 0.10 m   **(1 mark)**

(b) Determine the wavelength of the wave in the diagram.

> You may find this equation useful:
> $v = f \times \lambda$ or wave speed = frequency × wavelength

wavelength = ... m   **(1 mark)**

(c) Sketch a second wave on the diagram to show a higher amplitude and shorter wavelength.   **(2 marks)**

3 When a wave travels through a material the average position of the particles of the material remains constant. Explain how particles move when a sound wave travels through the air.

> Guided

When a sound wave moves through the air the particles ....

....

.... **(2 marks)**

Wave equations

1 Wave speed can be calculated if the distance travelled and the time taken are known.

(a) Write the units for the following quantities.

average speed = ...

distance = ...

time = ...   **(2 marks)**

(b) Give the equation that links these three quantities.

.... **(1 mark)**

2 Whales communicate over long distances by sending sound waves through the oceans. It takes 20 seconds for the sound waves to travel in sea water between two whales 30 kilometres apart.

Calculate the speed of sound in water in metres/second.

distance travelled by the waves (in metres) = ...

time taken = ...

....

average speed of sound = ... m/s   **(3 marks)**

3 A sound wave has a wavelength of 0.017 m and a frequency of 20 000 Hz.

(a) Write the equation that links wave speed, frequency and wavelength.

.... **(1 mark)**

(b) Calculate the speed of the wave.

wave speed = ... m/s   **(2 marks)**

4 An icicle is melting into a pool of water. Drops fall at a frequency of 2 Hz, producing small waves that travel across the water at 0.05 m/s. Calculate the wavelength of the small waves. State the unit.

> Remember to write down the equation you are using before you substitute values.

wavelength = ... unit .... **(3 marks)**

Measuring wave velocity

Guided

1 A tap is dripping into a bath. Three drops fall each second producing small waves that are 5 cm apart. Calculate the speed of the small waves across the water. State the units.

> Remember: frequency = number of waves per second and wavelength = distance between each wave.

frequency of the waves (f) = ...

wavelength of the waves (λ) = ...

speed of waves = ... unit .... **(3 marks)**

2 What is the wavelength of water waves with a frequency of 0.25 Hz travelling at a speed of 2 m/s?

☐ **A** 0.5 m   ☐ **C** 4 m

☐ **B** 0.125 m   ☐ **D** 8 m   **(1 mark)**

3 A bottlenose dolphin finds its food by echolocation. It emits sound waves which travel through the water at a speed of 1500 m/s and locates a fish 150 m away. Calculate the time taken for the dolphin to receive the echo from the fish.

> You will need to recall the equation:
> $$v = \frac{x}{t}$$

(3 marks)

4 Eliza and Charlie set up an experiment to estimate the speed of sound in air. They use a brick wall at school and stand 50 metres away. Charlie knocks two pieces of wood together and Eliza measures the time of the echo using a stopwatch.

Suggest **two** changes that Eliza and Charlie could make to their experiment to make their results more reliable.

....

....

....

....

.... **(2 marks)**

Had a go ☐ Nearly there ☐ Nailed it! ☐

Waves and boundaries

1 Identify the correct term for a wave changing direction at the boundary of materials of two different densities:

☐ **A** Diffraction

☐ **B** Dispersion

☐ **C** Reflection

☐ **D** Refraction **(1 mark)**

2 The diagram below shows light waves travelling from a less dense material into a denser material. Draw the features identified below on the wave diagram.

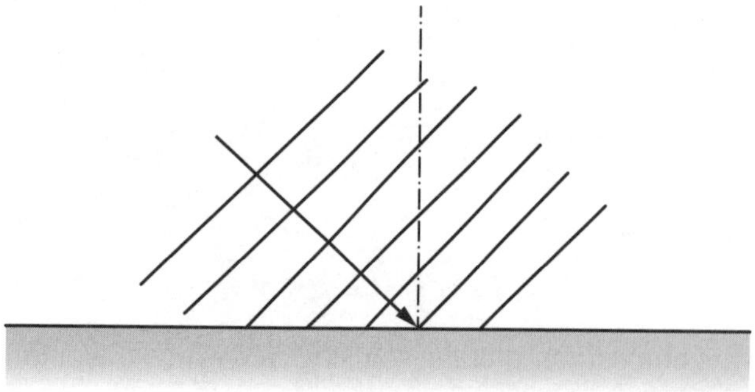

(a) The 'normal' line **(1 mark)**

(b) A ray to show the change of direction of the wave in the second material **(1 mark)**

(c) Wave fronts in the second material **(2 marks)**

Guided

3 A wave may be refracted when it passes from one material to another material at an angle to the normal. Describe how the direction of a wave changes when it passes:

(a) from one material to a different, less dense material

The wave bends .... the normal. **(1 mark)**

(b) from one material to a different, denser material.

The wave bends .... the normal. **(1 mark)**

4 Draw a ray diagram to represent a light wave passing from the air **through** glass and back to air, **perpendicular** to the boundary of the materials.

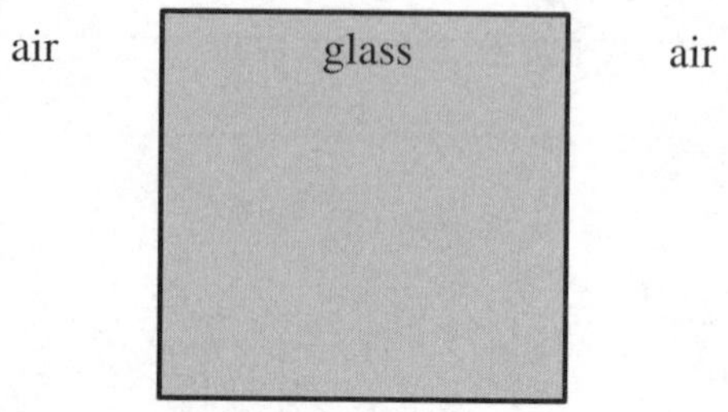

(2 marks)

Had a go ☐   Nearly there ☐   Nailed it! ☐

Waves in fluids

1   A ripple tank is used to investigate waves.

(a) Describe how a ripple tank may be used to measure the frequency of water waves.

....

....

....   **(2 marks)**

(b) Describe how to find the wavelength of the waves in the ripple tank.

....

....

....   **(2 marks)**

(c) State the equation you can use with the data collected in (a) and (b) to determine wave speed.

....   **(1 mark)**

(d) Identify the control variable when using a ripple tank to investigate wave speed.

....   **(1 mark)**

2   Describe a suitable conclusion for the ripple tank investigation referred to in Q1. Your conclusion should include two factors that should be moderated in this experiment.

> Guided

A ripple tank can be used to determine the three values for ....

....

....

....   **(3 marks)**

3   The ripple tank experiment uses several pieces of equipment. Complete the table below to describe the hazard associated with each component and suggest a measure to minimise the risk of harm.

> Identify the hazard and describe the safety measure for each mark.

Component	Hazard	Safety measure
water		
electricity		
strobe lamp		

(3 marks)

Extended response – Waves

A man has dropped his door key into the pond. The key appears to be in a different position because of this phenomenon shown in the diagram.

Explain why the key appears to be in a different position.

Your answer should identify the actual position of the key.

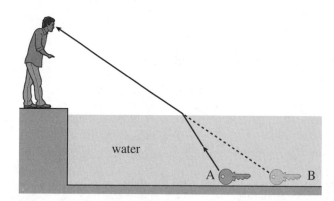

You will be more successful in extended response questions if you plan your answer before you start writing.

The question asks you to give a detailed explanation of what happens when a light wave passes through the boundary between two different materials. Think about:

- the property of waves that is illustrated
- the types of materials involved in the wave behaviour shown in the diagram
- the causes of the wave behaviour shown in the diagram
- why the driver believes the key is in a different position from the actual position.

You should try to use the information given in the question.

....

....

....

....

....

....

....

....

....

....

....

.... **(6 marks)**

Electromagnetic spectrum

1 Which statement about electromagnetic (EM) waves is correct?

☐ **A** EM waves are tranverse waves that travel at the same speed in a vacuum.

☐ **B** EM waves are transverse waves that have the same frequency in a vacuum.

☐ **C** EM waves are longitudinal waves that travel at the same speed in a vacuum.

☐ **D** EM waves are transverse waves that travel at different speeds in a vacuum. **(1 mark)**

2 Microwaves and ultraviolet are types of radiation. Identify the statement that describes these waves correctly.

☐ **A** Microwaves have a higher frequency than ultraviolet.

☐ **B** Microwaves and ultraviolet are transverse waves.

☐ **C** Microwaves have a shorter wavelength than ultraviolet.

☐ **D** Microwaves and ultraviolet are longitudinal waves. **(1 mark)**

3 The chart below represents the electromagnetic spectrum. Some types of electromagnetic radiation have been labelled.

longest wavelength/
lowest frequency   shortest wavelength/
                                          highest frequency

←— radio waves —→ ←— C —→ ←infrared→B   ←→←— A —→
                                    ultra-
                                    violet   ←gamma→
                                    rays (UV)   rays

(a) Name the three parts of the spectrum that have been replaced by letters in the diagram.

A: ....

B: ....

C: .... **(3 marks)**

(b) Explain why different parts of the electromagnetic spectrum have different properties.

.... **(1 mark)**

4 The speed of electromagnetic waves in a vacuum is 300 000 000 m/s. A radio wave has a wavelength of 240 m. Calculate the frequency of the radio wave.

frequency = ... Hz **(3 marks)**

 **Practical skills** # Investigating refraction

1 (a) Suggest a method that could be used to investigate the refraction of light using a glass block and a ray box.

....

....

....

....

....

....

.... **(5 marks)**

> **Guided** (b) Explain what conclusion you would expect to find using the method you have outlined in (a). Your answer should refer to the angle of incidence and the angle of refraction.

When a light ray travels from air into a glass block ....

....

.... **(2 marks)**

(c) Explain what would be observed if the light ray, travelling through the air, entered the glass at an angle of 90° to the surface of the glass.

.... **(1 mark)**

2 State **three** hazards associated with investigating reflection with a ray box and suggest safety measures that could be taken to minimise the risk.

....

....

....

....

.... **(3 marks)**

3 When investigating refraction, discuss the importance of transparent, translucent and opaque properties when selecting suitable materials to use with visible light.

> **Guided**

| Only two of the properties identified in the question are suitable for investigating visible light. |

Transparent materials allow .... and so are ....

for investigating refraction.

Translucent materials allow .... but ....

so measuring refraction ....

Opaque materials .... and so are ....

for investigating refraction.

                                                                   (3 marks)

Dangers and uses

1 (a) Identify the property of EM spectrum waves that, as it gets higher, increases potential danger to body cells.

.... **(1 mark)**

(b) Complete the table of EM spectrum waves to show the order of increasing potential danger to body cells by adding numbers 1, 2, 3 and 4 (1 being the lowest potential danger).

	Type of EM spectrum wave	Order of increasing potential danger
A	X-rays	
B	microwaves	
C	gamma-waves	
D	infrared waves	

(2 marks)

2 Identify the **two** correct uses of each type of electromagnetic radiation.

Remember that you need to choose **two** correct answers for each question (a), (b) and (c).

(a) **infrared** ☐ A night-vision goggles   ☐ B broadcasting TV programmes

☐ C TV remote control   ☐ D sun-tan lamps   **(1 mark)**

(b) **ultraviolet** ☐ A thermal imaging   ☐ B disinfecting water

☐ C cooking food   ☐ D security marking   **(1 mark)**

(c) **gamma-waves** ☐ A sterilising food   ☐ B communicating with satellites

☐ C security systems   ☐ D treating cancer   **(1 mark)**

3 Describe **three** uses of X-rays and suggest where each is used.

....

....

.... **(3 marks)**

4 Describe a danger of each of the following waves by completing each sentence using words from the box. Use each word only once.

eyes	heating	cells	cancer	thermal	burns

(a) Microwaves cause damage to

.... **(2 marks)**

(b) Infrared waves transfer

.... **(2 marks)**

(c) Ultraviolet waves can damage

.... **(2 marks)**

Changes and radiation

1 Which of the following statements about electrons is true?

☐ **A** Electrons only change orbit when they emit electromagnetic radiation.

☐ **B** Electrons only absorb electromagnetic radiation.

☐ **C** Electrons always change orbit when electromagnetic radiation is absorbed.

☐ **D** Electrons can only orbit the nucleus at defined energy levels within the atom. **(1 mark)**

2 Electromagnetic radiation can be absorbed by electrons that orbit an atomic nucleus.

⟩ **Guided** ⟩

(a) Describe what happens to an electron that absorbs electromagnetic radiation.

When an electron absorbs electromagnetic radiation ....

.... **(1 mark)**

(b) Describe what happens to an electron that emits electromagnetic radiation.

When an electron emits electromagnetic radiation ....

.... **(1 mark)**

3 The diagram shows different electron energy levels in an atom. With reference to the diagram, identify the correct statement that describes changes in the energy levels of an electron.

> Think how energy is absorbed or emitted for the electron to move between these levels (n).

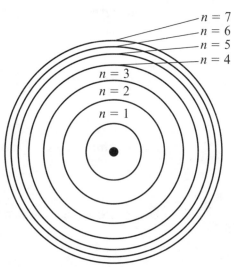

		Electron movement	
☐	**A**	$n = 3 \rightarrow n = 1$	Radiation is absorbed by the atom.
☐	**B**	$n = 1 \rightarrow n = 2$	Radiation is emitted by the atom.
☐	**C**	$n = 3 \rightarrow n = 1$	Radiation is emitted by the atom.
☐	**D**	$n = 3 \rightarrow n = 2$	Radiation is absorbed by the atom.

(1 mark)

Extended response – Light and the electromagnetic spectrum

X-rays and gamma-waves are widely used in a number of applications. Compare and contrast these waves and give examples of how they can be used safely in industry.

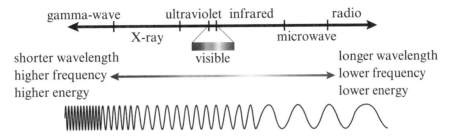

You will be more successful in extended response questions if you plan your answer before you start writing.

The question asks you to give a detailed explanation of the properties, uses and dangers of X-rays and gamma-waves. Think about:

- the types of waves and how you would describe them
- the dangers of both types of waves and the reasons why they can be dangerous
- examples of how the waves are used in medicine
- examples of how the waves are used in industry.

You should try to use the information given in the question and in the diagram.

....

....

....

....

....

....

....

....

....

....

....

....

....

....

.... (6 marks)

Structure of the atom

1 Which statement about the nucleus of an atom is **incorrect**?

☐ **A** The nucleus contains more than 99% of the mass of an atom.

☐ **B** The nuclei of atoms of the same element may contain different numbers of protons.

☐ **C** The nuclei of atoms of the same element may contain different numbers of neutrons.

☐ **D** A neutral atom will have the same number of electrons as protons.   **(1 mark)**

2 Complete the diagram to show the location and charge of:

(a) protons   **(1 mark)**

(b) neutrons   **(1 mark)**

(c) electrons.   **(1 mark)**

> Recall what you know about the Rutherford and Bohr models of the atom.

3 (a) Explain why neutral atoms have no overall charge in terms of their particles.

> Guided

The number of ....

is equal to ....

.... **(2 marks)**

(b) State what will happen to the overall charge if an atom loses an electron.

.... **(1 mark)**

4 (a) State what is meant by the term molecule.

.... **(1 mark)**

(b) Give an example of the following at room temperature:

(i) a molecule of liquid

> Remember that molecules are given a chemical formula.

.... **(1 mark)**

(ii) a molecule of a gaseous element

.... **(1 mark)**

(iii) a molecule of a gaseous compound.

.... **(1 mark)**

Atoms and isotopes

1 State what is meant by each term:

> **Guided**

    (a) nucleon: *the name given to particles in the* .... **(1 mark)**

    (b) atomic number: .... **(1 mark)**

    (c) mass number: .... **(1 mark)**

2 (a) State what all atoms of the same element have in common in their nuclei.

    .... **(1 mark)**

    (b) Explain what is meant by the term isotope.

    ....

    .... **(2 marks)**

3 $_3^6Li$, $_3^7Li$, and $_3^8Li$ are all isotopes of lithium. Identify the incorrect statement.

    ☐ **A** The mass numbers of the atoms are different.

    ☐ **B** The proton numbers of the atoms are different.

    ☐ **C** The atomic numbers of the atoms are the same.

    ☐ **D** The neutron numbers of the atoms are different. **(1 mark)**

4 Explain why different isotopes of the same element will still be neutrally charged even though the nucleon number is different.

> **Guided**

    Isotopes will be neutral because ....

    ....

    ....

    .... **(2 marks)**

5 Complete the table below to show the components of isotopes. The first one has been done for you.

> Remember that the number of neutrons = mass number – atomic number.

Description of atom	Symbol	Protons	Neutrons
potassium (K) (atomic number 19; mass number 39)	$_{19}^{39}K$	19	20
carbon (C) (atomic number 6; mass number 12)			
chlorine (Cl) (atomic number 17; mass number 35)			
sodium (Na) (atomic number 11; mass number 23)			

(3 marks)

Atoms, electrons and ions

1 The diagram shows a neutral atom of boron.

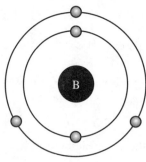

(a) State how many protons are in the nucleus of the atom.

.... **(1 mark)**

(b)   Explain how you have reached your answer to (a).

....

.... **(2 marks)**

2 Select the correct statement about atoms.

☐ **A** Electrons orbit at fixed distances from the nucleus.

☐ **B** Electrons orbit the nucleus at random distances from the nucleus.

☐ **C** An electron can be lost from the atom when it emits electromagnetic radiation.

☐ **D** Electrons move to a lower orbit when they absorb electromagnetic radiation. **(1 mark)**

3 Describe what happens to an electron when its atom:

> **Guided**

(a) absorbs electromagnetic radiation

When an atom absorbs electromagnetic radiation ....

.... **(2 marks)**

(b) emits electromagnetic radiation.

When an atom emits electromagnetic radiation ....

.... **(2 marks)**

4 (a) Write these in the correct part of the table.

F^- Li Na^+ B^+ K^+ Cu

Atoms	Ions

(2 marks)

(b) Explain your choices.

> Describe the influence of electrons on both atoms and ions.

....

.... **(2 marks)**

Ionising radiation

1 Describe the structure of alpha, beta and gamma radiation.

Guided

(a) An alpha particle is ....

.... **(2 marks)**

(b) A beta particle is an...

and has a charge .... **(2 marks)**

(c) A gamma wave is ....

.... **(2 marks)**

2 Select the correct description of an alpha particle.

☐ **A** helium nucleus with charge −2

☐ **B** helium nucleus with charge +2

☐ **C** high-energy neutron

☐ **D** ionising electron   **(1 mark)**

3 Match the types of radiation with the correct penetrating power.

Type of radiation		Penetrating power
alpha		low, stopped by thin aluminium
beta minus		very high, stopped by very thick lead
gamma		very low, stopped by 10 cm of air

(3 marks)

4 Identify the type of radiation that would be emitted in each decay.

> Remember the law of conservation of matter.

(a) carbon-10 (6 protons, 4 neutrons) → boron-10 (5 protons, 5 neutrons)

.... **(1 mark)**

(b) uranium-238 (92 protons, 146 neutrons) → thorium-234 (90 protons, 144 neutrons)

.... **(1 mark)**

(c) helium-5 (2 protons, 3 neutrons) → helium-4 (2 protons, 2 neutrons)

.... **(1 mark)**

5 State why it is **not** possible to determine exactly when any nucleus will decay.

.... **(1 mark)**

Background radiation

1 The pie chart shows the sources of background radiation; 50% of this comes from the element radon.

> Think about where radon originally comes from.

(a) Describe what radon is and how it occurs.

....

.... **(2 marks)**

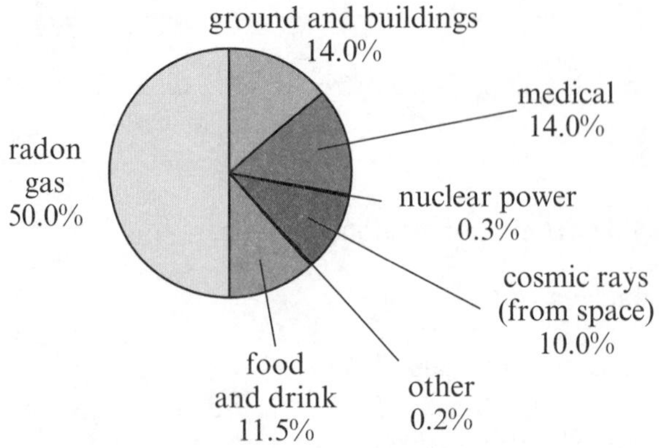

ground and buildings
14.0%

medical
14.0%

nuclear power
0.3%

cosmic rays
(from space)
10.0%

other
0.2%

food
and drink
11.5%

radon
gas
50.0%

2 Give **two** reasons why radon levels can vary across the UK.

> Guided

Levels can vary because of the different rocks ....

They can also vary ....

.... **(2 marks)**

3 Complete the table by giving **two** examples each of natural and man-made sources of background radiation.

Sources of background radiation	
Natural	**Man-made**

> Use the pie chart to help you.

(2 marks)

4 A scientist in the south-east of England measures the background radiation count three times. Her colleague conducts the same experiment in the south-west. Their results are shown in the table below.

Test number	1	2	3	Average
south-east activity (Bq)	0.30	0.24	0.27	
south-west activity (Bq)	0.31	0.28	0.32	

(a) Calculate the average activity for each sample and write it in the table. **(2 marks)**

(b) State which area has the highest level of background radiation.

.... **(1 mark)**

5 Describe how radon gas can get into homes and buildings.

....

.... **(2 marks)**

Measuring radioactivity

1 In 1896 Henri Becquerel discovered that uranium salts led to a darkening of photographic film over time. Today nuclear industry workers wear safety badges, containing photographic film. Explain why these badges are worn.

Badges containing photographic film monitor levels of ....

.... **(1 mark)**

2 Put the steps in the boxes below into the correct order to show how a Geiger–Muller tube detects nuclear radiation.

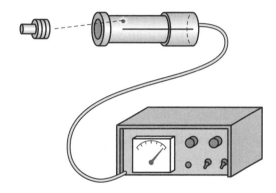

A Atoms of argon are ionised.	→	B The amount of radiation detected is shown by the rate meter.	→	C A thin wire is connected to +400V.	→	D Electrons travel towards the thin wire.

	→		→		→	

(3 marks)

3 A student claims that the more ionising the radiation, the more effective the G–M tube is at detecting levels of radiation. Explain whether the student is correct or not.

....

....

....

....

....

.... **(3 marks)**

Models of the atom

1   (a)   What particle did Niels Bohr's model of the atom specifically develop new ideas for?

     ☐   **A**   the electron   ☐   **C**   the neutron

     ☐   **B**   the proton   ☐   **D**   the nucleus   **(1 mark)**

   (b)   Explain how the Bohr model of the atom improved on Rutherford's model.

     ....

     ....

     ....

     ....

     ....

     .... **(4 marks)**

2   Describe the evidence that enabled Rutherford to make his claim about the nucleus.

> Rutherford used the 'gold-leaf' experiment to gather this evidence.

     ....

     ....

     ....

     ....

     ....

     .... **(3 marks)**

3   Compare and contrast the plum pudding and Rutherford models of the atom.

> Use the diagrams to help you.

Guided

negative electron 'plums'

positive 'pudding'

neutron — nucleus — proton — electron

The plum pudding model showed the atom as ....

....

....

whereas the Rutherford model showed the atom as ....

....

.... **(4 marks)**

Beta decay

1 State the nature of the two types of beta particle.

β^+ particle ....

β^- particle .... **(2 marks)**

2 Draw lines to link the boxes to complete the sentences about beta-minus and beta-plus decay. One has been done for you.

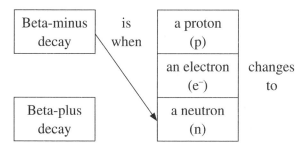

Beta-minus decay	is when	a proton (p)		an electron (e^-)		a high-energy positron (e^+).
		an electron (e^-)	changes to	a proton (p)	releasing	a high-energy electron (e^-).
Beta-plus decay		a neutron (n)		a neutron (n)		a slow neutron (n).

(2 marks)

3 Using the data in the table, complete the equations below.

7 Li lithium 3	9 Be beryllium 4	11 B boron 5	12 C carbon 6	14 N nitrogen 7	16 O oxygen 8	19 F fluorine 9	20 Ne neon 10
23 Na sodium 11	24 Mg magnesium 12	27 Al aluminium 13	28 Si silicon 14	29 P phosphorus 15	31 S sulfur 15	35.5 Cl chlorine 17	40 Ar argon 18

(a) $^{14}_{6}C \rightarrow \:^{14}_{\_\_}N + \:^{0}_{-1}e$ **(1 mark)**

(b) $^{23}_{\_\_}Mg \rightarrow \:^{23}_{11}Na + \:^{0}_{1}e$ **(1 mark)**

4 Describe what happens in:

> **Guided**

(a) beta-minus decay

> In beta decay neutrons and protons undergo changes producing high-energy beta particles. Describe these changes and the particles emitted.

In β^- decay, an n ....

....

.... **(2 marks)**

(b) beta-plus decay.

In β^+ decay, a p ....

....

.... **(2 marks)**

Radioactive decay

1 Radium-222 undergoes alpha decay. Identify which **two** of the following statements are true.

☐ **A** The positive charge of the nucleus is reduced by 4.

☐ **B** The mass number is reduced by 4.

☐ **C** The atomic number is reduced by 2.

☐ **D** The nucleus gains an extra proton.   **(1 mark)**

2 Beta decay has two forms. Name the two types of beta decay and give the charge for each type.

> **Guided**

beta- .... , charge ....

beta- .... , charge .... **(2 marks)**

3 A student claims that, when alpha radiation is given out, there is no change in the mass of the nucleus.

Explain if the student is right or wrong.

....

.... **(2 marks)**

4 Describe what happens to the mass **and** charge of the nucleus in neutron decay.

....

....

.... **(2 marks)**

5 Identify the correct term that could be used in a description of gamma decay.

☐ **A** electron   ☐ **C** positron

☐ **B** photon   ☐ **D** proton   **(1 mark)**

6 (a) Complete each equation and state what type of decay is shown.

> Check that the A and Z numbers obey the conservation laws.

(i) $_{84}^{\ \ }\text{Po} \rightarrow {}_{2}^{4}\text{He} + {}_{82}^{204}\text{Pb}$ type of decay .... **(2 marks)**

(ii) $_{\ \ }^{222}\text{Rn} \rightarrow {}_{2}^{4}\text{He} + {}_{84}^{218}\text{Po}$ type of decay .... **(2 marks)**

(iii) $_{19}^{42}\text{K} \rightarrow {}_{-1}^{0}\text{e} + {}_{20}^{\ \ }\text{Ca}$ type of decay .... **(2 marks)**

(iv) $_{4}^{\ \ }\text{Be} \rightarrow {}_{0}^{1}\text{n} + {}_{4}^{8}\text{Be}$ type of decay .... **(2 marks)**

(b) Nuclear decay results in a loss of energy from the nucleus. State the reason for this and the form of the emitted energy.

....

....

.... **(2 marks)**

Had a go ☐   Nearly there ☐   Nailed it! ☐

Half-life

1 State what is meant by the term half-life.

....

.... **(2 marks)**

2 A sample of thallium-208 contains 16 million atoms. Thallium-208 has a half-life of 3.1 minutes.

(a) State the number of nuclei that will have decayed in 3.1 minutes.

number of nuclei = ... **(1 mark)**

(b) Calculate the number of unstable thallium nuclei left after 9.3 minutes.

number of unstable thallium nuclei left = ... **(2 marks)**

3 A student measured the activity of a radioactive sample for 30 minutes. She plotted the graph of activity against time, shown on the right.

Guided

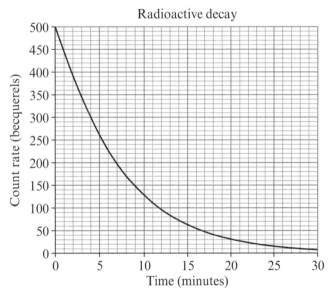

Radioactive decay

Use the graph to calculate the half-life of the sample.

> You could take any point on the line as a starting point for calculating half-life.

The activity is .... Bq at .... min.

Half this activity is .... Bq, which is at .... min

so the half-life is ....

half-life = ... min **(3 marks)**

Dangers of radiation

1   The hazard symbol shown is used to warn that sources of ionising radiation may be present.

> Think about the use of ionising radiation in hospitals, other medical centres and in industry.

Give **two** places where this symbol may be displayed.

....

.... **(2 marks)**

2   Which is the most damaging radiation by contamination?

☐   **A**   alpha radiation

☐   **B**   beta-negative radiation

☐   **C**   beta-positive radiation

☐   **D**   gamma radiation   **(1 mark)**

3   (a)   State what is meant by the term ionising.

> **Guided**

Ionising means to convert .... **(1 mark)**

   (b)   Explain why ions are dangerous in the body.

Ions in the body can cause ....

which can lead to .... **(2 marks)**

4   (a)   Describe how employers using radioactive sources can take steps to reduce the exposure of their workers to ionising radiation.

....

....

....

....

.... **(3 marks)**

   (b)   When risk of exposure to radiation has been minimised through procedure, describe how workers can be monitored to further improve their safety.

> Photographic film is an important tool in monitoring levels of exposure to radiation. Think how this is used.

....

....

.... **(2 marks)**

Contamination and irradiation

1 During the First World War (1914–18) soldiers and airmen were issued with watches that had hands and numbers that glowed in the dark. The hands and numbers had been painted with luminous paint that contained radium. Radium was discovered in 1898 and found to be radioactive. In the 1920s many of the workers who painted the watches became very ill.

Discuss why it was **not** banned from being used on watches until the 1920s.

> You may need to make an assumption about the extent of scientific research into radium before 1920.

Before 1920 the harmful effects of radioactivity ....

so it was thought that ....

It was banned ....

.... **(3 marks)**

2 Draw a line from each term to its correct description.

Term	Description
external contamination	A radioactive source is eaten, drunk or inhaled.
internal contamination	A person becomes exposed to an external source of ionising radiation.
irradiation	Radioactive particles come into contact with skin, hair or clothing.

(2 marks)

3 Give an example of how a person may be subjected to:

(a) external contamination

.... **(1 mark)**

(b) internal contamination.

.... **(1 mark)**

4 Ionising radiation can damage or kill body cells. Explain why alpha particles are more dangerous from a source of internal contamination than from a source of irradiation.

> Alpha particles can only travel short distances before they collide with another particle and lose their energy. This can have serious consequences near to the body.

Internal contamination means that the alpha particles ....

.... with the body through ....

where they are likely to cause ....

Irradiation by alpha particles is less likely ....

because they .... and are therefore

.... **(4 marks)**

Extended response – Radioactivity

1   Ionising radiation will travel through some materials but will be stopped by others. The diagram shows three materials and how well they absorb different types of radiation.

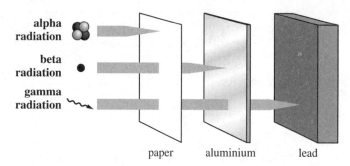

alpha radiation
beta radiation
gamma radiation

paper   aluminium   lead

Explain the different abilities of alpha, beta and gamma radiations to penetrate materials and to ionise substances. Use the diagram to help you.

> You will be more successful in extended response questions if you plan your answer before you start writing. Think about the following:
>
> • which radiation is the least penetrating and which is the most penetrating
>
> • which materials absorb the different types of radiation
>
> • what happens when a substance is ionised by radiation
>
> • which radiation is the least ionising and which is the most ionising
>
> • how much energy is transferred when radiation causes ionisation
>
> • how this is related to the penetrating ability of the radiation.
>
> You should try to use the information given in the question.

....

....

....

....

....

....

....

....

....

....

....

.... (6 marks)

Work, energy and power

1 A kettle has a power rating of 2500 W. How much energy does it transfer in 5 seconds?

   ☐ **A** 2500 J

   ☐ **B** 500 J

   ☐ **C** 500 W

   ☐ **D** 12 500 J   **(1 mark)**

2 Give the energy store that increases in each of these examples:

   > Energy transfers result in the movement of energy from one store to another.

   (a) when a mass is lifted through a height

   .... **(1 mark)**

   (b) when a pan of water at 20 °C is heated to 70 °C

   .... **(1 mark)**

   (c) when an extra cell is added to a circuit

   .... **(1 mark)**

3 A microwave cooker heats a drink in 20 seconds using 15 000 J of electrical energy. Calculate the power of the microwave cooker. State the unit.

> **Guided**

energy transferred = ... J, time taken = ... s

$P = \dfrac{E}{t} =$

                                                      power = ... unit .... **(3 marks)**

4 A student weighing 600 N climbs 20 stairs to a physics lab. Each stair is 0.08 m high. Calculate the work done by her muscles to climb the stairs. State the unit.

                                              work done = ... unit .... **(3 marks)**

5 A student watches a programme on his television, which has a power rating of 200 W and uses 360 000 J of energy during the viewing. Calculate the time the student spends watching the television.

                                              time taken = ... s **(3 marks)**

Extended response – Energy and forces

Wind turbines are designed to use the kinetic energy of moving air to turn the turbine blades.

Use a Sankey diagram to help you to explain how every 100 J of kinetic energy from the wind can be transferred to the useful kinetic energy store in the turbine. The transfer is 35% efficient with some energy transferred to the thermal store of the turbine (and environment) and to the sound energy store of the turbine.

Describe the process and how the diagram illustrates this.

> You will be more successful in extended response questions if you plan your answer before you start writing.
>
> You should try to use the information given in the question.
>
> The question asks you to give a detailed explanation of the process of energy transfers from air to turbine blades and to draw a Sankey diagram as part of your answer. Think about:
>
> - the source of the energy transfers
> - the useful and wasted energy in the process
> - how efficiency can be calculated and improved
> - the components and structure of a Sankey diagram
> - the consequences of a mechanical process.

....

....

....

....

....

....

....

....

.... **(6 marks)**

Interacting forces

1 (a) Give the **three** types of fields that cause objects to interact with each other **without** making contact.

.... **(3 marks)**

> **Guided**

(b) Explain which of these is different from the other two and why.

> Two of these fields have opposite poles or charges but one acts in only one direction.

A .... is different because it only...................................

whereas both..

.... **(2 marks)**

2 Which **two** of the following correctly describe the similarities between magnetic and electrostatic fields?

   ☐ **A** Like poles/charges repel.

   ☐ **B** Like poles/charges attract.

> Think about the effects of the poles and charges.

   ☐ **C** Opposite poles/charges attract.

   ☐ **D** Opposite poles/charges repel.   **(2 marks)**

3 Explain why weight and normal contact force are described as vectors.

> **Guided**

Weight is a vector because ....

....

Normal contact force is a vector because ....

.... **(2 marks)**

4 A student moves a luggage bag forwards, as shown in the diagram.

(a) Name the forces for the horizontal motion and state which force is larger.

....

....

.... **(2 marks)**

(b) Name the balanced contact forces in the vertical direction.

.... **(1 mark)**

Circuit symbols

1 Explain why electrical circuits are drawn using agreed symbols and conventions.

.... **(1 mark)**

2 (a) Identify the two components that are shown by the circuit symbols.

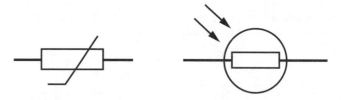

☐ **A** thermistor and LED   ☐ **C** thermistor and LDR

☐ **B** resistor and diode   ☐ **D** variable resistor and LDR   **(1 mark)**

> **Guided**

(b) Describe how the components you have chosen respond automatically to changes in the environment.

(i) The .... responds by ....

.... **(1 mark)**

(ii) The .... responds by ....

.... **(1 mark)**

3 Complete the table of circuit symbols below. Four of the answers are given for you.

Component	Symbol	Purpose
ammeter		
		provides a fixed resistance to the flow of current
	⊣▷⊢	
		allows the current to be switched on or off

(4 marks)

Series and parallel circuits

1 (a) Each lamp in these circuits is identical. Write the current for each ammeter on the circuit diagrams.

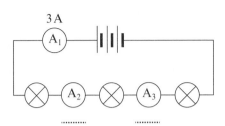

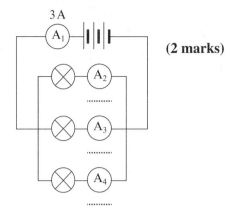

(2 marks)

(b) Explain the rules for current in series and parallel circuits.

In a series circuit the current ....

In a parallel circuit the current .... **(2 marks)**

2 (a) Each lamp in these circuits is identical. Write the potential difference for each voltmeter in the circuit diagrams.

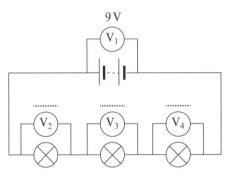

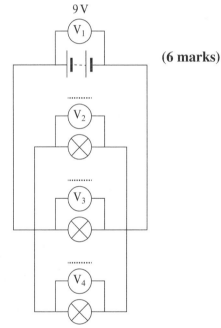

(6 marks)

(b) Explain the rules for potential difference in series and parallel circuits.

In a series circuit the potential difference ....

....

In a parallel circuit the potential difference ....

.... **(2 marks)**

3 Which statement about voltmeters and ammeters is correct?

    ☐ **A** Ammeters are always connected in series in both series and parallel circuits.

    ☐ **B** Voltmeters are always connected in series in both Saries and parallel circuits.

    ☐ **C** Voltmeters are connected in parallel in series circuits and in series in parallel circuits.

    ☐ **D** Ammeters are connected in series in series circuits and in parallel in parallel circuits.

(1 mark)

Current and charge

1 The electric current flowing in a circuit is 4 A.

   (a) Explain what is meant by an electric current.

   An electric current is the rate ....

   .... **(2 marks)**

   (b) The current flows for 8 seconds. Calculate how much charge has flowed. Give the unit.

   > You may find this equation useful:
   > $Q = I \times t$

   charge = ... unit .... **(3 marks)**

2 The diagram shows a series circuit.

   (a) Give the reading on ammeter:

   (i) A_1 .... A **(1 mark)**

   (ii) A_3 .... A **(1 mark)**

   0.3 A

   (b) State how you could increase the size of the current flowing through the circuit.

   .... **(1 mark)**

   (c) Explain why the current measured by ammeter A_2 is the same as A_1 and A_3.

   The electrons move around the ....

   so the current leaving the cell is the same as ....

   returning to it. **(2 marks)**

3 A student is investigating how current carries charge around the circuit.

   (a) Draw a circuit diagram to show the measuring instrument needed in the circuit in order to calculate charge.

   (2 marks)

   (b) State what else the student would need to use to collect enough data in order to calculate charge.

   .... **(1 mark)**

Energy and charge

1 State what is meant by current and potential difference. Include the word 'charge' in your answers.

Current is the ....

Potential difference is the .... **(2 marks)**

2 Calculate the amount of energy transferred to a 9 V lamp when a charge of 30 C is supplied.

charge = ... C

potential difference = ... V

so E = ...

You may find this equation useful:

$E = Q \times V$

energy transferred = ... J   **(3 marks)**

3 Calculate the charge needed to transfer 125 J of energy to a string of fairy lights with a total potential difference of 5 V.

In questions like these, you will need to rearrange familiar equations.

In this case, $E = Q \times V$ will become $Q = E \div V$

charge = ... C   **(3 marks)**

4 Calculate the energy transferred by a resistor when there is a current of 0.15 A through the resistor for 200 s. The potential difference across the resistor is 20 V.

This question will need a two-step answer. Step one, calculate the charge using $Q = I \times t$. Step two, calculate the energy transferred using $E = Q \times V$.

energy transferred = ... J   **(3 marks)**

Ohm's law

1 Which quantity is the ohm (Ω) a unit of?

☐ **A** current ☐ **C** potential difference

☐ **B** energy ☐ **D** resistance

(1 mark)

2 Explain what Ohm's law means.

> **Guided**

Ohm's law means that ....

....

is directly proportional ....

.... **(2 marks)**

3 Use Ohm's law to calculate the resistance of each resistor:

> You may find this equation useful: $R = V \div I$

(a) a resistor with a potential difference of 12 V across it and a current of 0.20 A passing through it

resistance = ... Ω **(2 marks)**

(b) a resistor with a potential difference of 22 V across it and a current of 0.40 A passing through it

resistance = ... Ω **(2 marks)**

(c) a resistor with a potential difference of 9 V across it and a current of 0.03 A passing through it

resistance = ... Ω **(2 marks)**

(d) Identify which resistor has the highest resistance.

.... **(1 mark)**

4

(a) Sketch two lines on the graph to show two ohmic conductors of different resistances labelled A and B. **(2 marks)**

(b) From your graph, identify which line represents the resistor with the higher resistance.

.... **(1 mark)**

Resistors

1 The diagrams show resistors in series and resistors in parallel. The resistors in both circuits are identical.

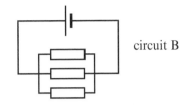

Series circuits: $R_T = R_1 + R_2 + R_3$
Parallel circuits: $1/R_T = 1/R_1 + 1/R_2 + 1/R_3$

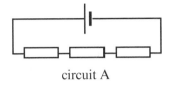

circuit A

circuit B

Identify the correct description of the total resistance in circuit B.

☐ **A** It is lower than the total resistance of circuit A.

☐ **C** It is the same as the total resistance of circuit A.

☐ **B** It is higher than the total resistance of circuit A.

☐ **D** The resistance is variable.   **(1 mark)**

2 A series circuit includes two 10 Ω resistors with a current of 2 A.

Recall the rule for current in a series circuit.

(a) Calculate the potential difference across each resistor.

potential difference = ... V   **(2 marks)**

(b) Calculate the total resistance of the circuit.

total resistance = ... Ω   **(2 marks)**

3 The diagram below shows a 20 Ω resistor, a 30 Ω resistor and a 150 Ω resistor connected in series with identical cells. The current measured by the ammeter is 0.03 A.

(a) Calculate the total resistance of the circuit.

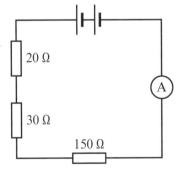

20 Ω

30 Ω

150 Ω

A

total resistance = ... Ω   **(2 marks)**

(b) (i) State the rule for potential difference in this type of circuit.

....

.... **(1 mark)**

(ii) Calculate the potential difference across each cell.

Step 1: Calculate total potential difference of the circuit using $V = I \times R$

where I = ... A   and R = (your answer from (a)) .... Ω

So V = ... = ... V

Step 2: Divide the answer at Step 1 by the number of cells in the series circuit.

potential difference = ... V   **(3 marks)**

207

I–V graphs

1 The graphs below show three types of component.

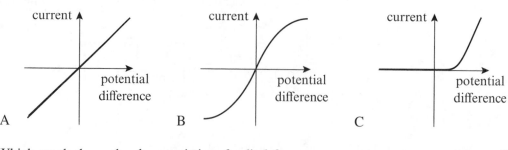

(a) Which graph shows the characteristics of a diode? .... **(1 mark)**

(b) Describe what happens to the current through the component shown in graph A as the potential difference increases.

> Remember to look at the axes labels, as well as the line, to work out how one variable is changing with the other.

....

.... **(2 marks)**

(c) Describe what happens to the current through the component shown in graph B as the potential difference increases.

....

.... **(2 marks)**

2 (a) Complete the I–V graphs for a fixed resistor and a filament lamp.   **(2 marks)**

Fixed resistor   Filament lamp

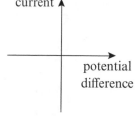   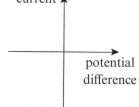

(b) Explain why the filament lamp graph has a different shape to the fixed resistor graph (at constant temperature).

> A fixed resistor (at constant temperature) obeys Ohm's Law but a filament lamp does not.

....

....

.... **(2 marks)**

3 Describe an experiment to collect data to enable the calculation of resistance.

Use an ammeter to measure current and a ....

A ....should be included to allow

....

Calculate resistance from .... **(5 marks)**

Electrical circuits

1 Electrical circuits can be connected in series or in parallel. Draw a circuit diagram below to show how you could measure the current and potential difference of two resistors connected in a:

(a) series circuit   **(3 marks)**

> Remember to include the ammeters and voltmeters in your diagrams.

(b) parallel circuit.   **(3 marks)**

(c) Describe the differences in the current and potential difference in a series and a parallel circuit.

....

....

....

....

.... **(4 marks)**

2 (a) Describe how a circuit may be set up to investigate the relationship between current and potential difference using a filament lamp.

....

....

....

....

.... **(3 marks)**

(b) State the law that may be applied to the data collected in (a) to find resistance.

.... **(1 mark)**

(c) Describe how resistance may be found using the data in (a) and a graphical method.

....

....

.... **(2 marks)**

Had a go ☐   Nearly there ☐   Nailed it! ☐

The LDR and the thermistor

1 Draw the circuit symbols for the components in the boxes provided.

Light-dependent resistor (LDR)	Thermistor

(2 marks)

2 Identify which variable will affect the resistance of a thermistor.

☐ **A** current

☐ **B** humidity

☐ **C** light

☐ **D** temperature

(1 mark)

3 Each sketch graph below shows the relationship between two variables.

> Recall independent and dependent variables.

(a) Describe how resistance changes with light.

.... **(1 mark)**

(b) Describe how resistance changes with temperature.

.... **(1 mark)**

4 An electric circuit in a car has a lamp connected in series with the battery and a thermistor. The lamp will only light up when the current is above a certain value. Describe the condition necessary for the lamp to light up.

Guided

The lamp lights up when the temperature is .... because the current

through the lamp and the thermistor ....

.... **(2 marks)**

5 The diagram shows a circuit with a light-dependent resistor. State what happens in terms of resistance and current when the level of light increases.

$V_{in} = 12\,V$

R_{top}
$1\,k\Omega$

LDR

V_{out}

$0\,V$

....

....

....

....

(2 marks)

Current heating effect

1 Which of the following appliances wastes energy due to the heating effect of a current?

    ☐ **A** lamp   ☐ **C** electric fire

    ☐ **B** kettle   ☐ **D** toaster   **(1 mark)**

2 Give **three** examples of how the heating effect of a current can be used in the home.

....

....

.... **(3 marks)**

3 Explain how the heating effect occurs in a conductor when a potential difference is applied. Include the words in the box.

> **Guided**

electrons   metal ions

lattice	ions	electrons	collisions

When a conductor is connected to a potential difference the free electrons move through the lattice of metal ions. As they do so ....

....

....

....

.... **(4 marks)**

4 A student claims that it is safe to use his computer, electric fan heater, desk lamp, coffee maker and toaster from the multi-socket extension lead plugged into a single socket, as all the plugs are earthed. Explain what danger the student is risking.

> Appliances that heat things up use a lot of current. Think about the heating effect when there is a high current.

....

....

....

....

.... **(4 marks)**

Had a go ☐ Nearly there ☐ Nailed it! ☐

Energy and power

1 Which quantity is **not** used to calculate the amount of energy transferred by an electrical device?

☐ **A** current

☐ **B** force

☐ **C** potential difference

☐ **D** time   **(1 mark)**

2 (a) A hotplate is used to heat a saucepan of water. The hotplate uses mains voltage of 230 V. The electric current through the hotplate is 5 A. Calculate the power of the hotplate in watts.

Guided

using the equation for power $P =$ ....

....

power = ... W   **(2 marks)**

(b) A mobile phone has a battery that produces a potential difference of 4 V. When making a call it uses a current of 0.2 A. A student makes a call lasting 30 seconds. Calculate the energy transferred by the mobile phone while the call is made. State the unit.

> You may find the equation $E = I \times V \times t$ useful.

energy transferred = ... unit ....   **(3 marks)**

3 The potential difference across a cell is 6 V. The cell delivers 3 W of power to a filament lamp.

(a) Calculate the current flowing through the lamp.

> Rearrange the equation linking power, current and potential difference ($P = I \times V$).

current = ... A   **(3 marks)**

(b) Calculate how much electrical energy is transferred to heat and light in the filament of the lamp when it is switched on for 5 minutes. State the unit.

energy transferred = ... unit ....   **(3 marks)**

a.c. and d.c. circuits

Guided

1 Circuits can operate using either an alternating current or a direct current.

(a) Explain what is meant by an **alternating** current.

Alternating current is an electric current that ....

.... **(2 marks)**

(b) Explain what is meant by a **direct** current.

Direct current is an electric current in which ....

.... **(2 marks)**

2 What is the potential difference and frequency of the mains supply in the UK?

☐ **A** 120 V and 50 Hz   ☐ **C** 50 V and 230 Hz

☐ **B** 230 V and 20 Hz   ☐ **D** 230 V and 50 Hz   **(1 mark)**

3 Calculate the energy transferred
for each of the following appliances:

> Remember to convert units where appropriate.

(a) a fan heater (2000 W) running for 15 minutes

energy transferred = ... J   **(2 marks)**

(b) a coffee maker (1.5 kW) running for 25 seconds

energy transferred = ... J   **(2 marks)**

(c) a tablet charger (10 W) running for 6 hours.

energy transferred = ... J   **(2 marks)**

Mains electricity and the plug

1 (a) Add labels to complete the diagram of a household plug. **(4 marks)**

....
(green and yellow)

....
(brown)

....

....
(blue)

outer insulation

cable grip

(b) Explain which wire the fuse is connected to.

....

.... **(2 marks)**

2 Draw a line from each wire to its correct function.

Wire	Function
brown	Electrical current leaves the appliance at 0 V through this wire.
blue	Electrical current enters the appliance at 230 V.
green/yellow	This is a safety feature connected to the metal casing of the appliance.

(2 marks)

3 Explain how a fuse in a plug works.

Guided

When a large current enters the live wire this transfers ....

....

....

....

The circuit is then .... **(4 marks)**

4 (a) Mains electricity supplies are fitted with a circuit breaker. Explain how a magnetic circuit breaker works.

> Magnetic circuit breakers rely on a high current producing a strong magnetic field.

....

....

....

.... **(3 marks)**

Guided

(b) Describe how the earth wire in a plug protects the user if the live wire becomes loose.

The earth wire is connected to the ....

If the live wire becomes loose ....

the current ....

instead of .... **(3 marks)**

Extended response – Electricity and circuits

Explain how a circuit can be used to investigate the change in resistance for a thermistor and a light-dependent resistor. Your answer should include a use for each component.

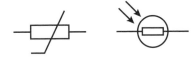

> You will be more successful in extended response questions if you plan your answer before you start writing.
>
> The question asks you to give a detailed explanation of how resistance changes in two types of variable resistor. Think about:
>
> - how resistance in a resistor can be measured and calculated
> - the variable that causes a change in resistance in a thermistor
> - the variable that causes a change in resistance in a light-dependent resistor
> - the consequence to the circuit of a change in resistance in a component
> - uses for thermistors and light-dependent resistors.
>
> You should try to use the information given in the question.

....

....

....

....

....

....

....

....

....

....

....

....

....

....

....

....

.... **(6 marks)**

Magnets and magnetic fields

1 Complete the diagrams below to show magnetic field lines for:

(a) a bar magnet   **(4 marks)**

> Remember to add arrows to the field lines.

(b) a uniform field.   **(2 marks)**

2 Give **two** similarities between a bar magnet and the magnetism of the Earth.

> **Guided**

Both a bar magnet and the Earth have ....

They also both have similar .... **(2 marks)**

3 An electric doorbell uses a temporary magnet to move the hammer, and ring the bell, when the button (switch) is pressed. It uses a coil with a soft iron core.

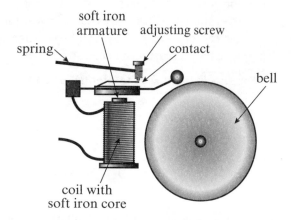

Explain why a temporary magnet rather than a permanent magnet is used for this application.

....

....

....

.... **(3 marks)**

4 Rajesh carries out an experiment to test for magnetic or non-magnetic materials by moving a permanent magnet near a range of small objects. Some of the objects are attracted to the magnet. Devise a second test that Rajesh could do to further sort the magnetic objects into temporary and permanent magnets using the same permanent magnet.

> Think of the difference between a temporary and a permanent magnet.

....

....

.... **(3 marks)**

Current and magnetism

1 The diagram shows the cross-sections of two current-carrying wires. The **cross** represents the conventional current moving into the page and the **dot** represents the conventional current moving out of the page.

You can use the 'right-hand grip' rule to help you answer this question.

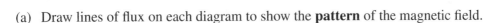

    (a) Draw lines of flux on each diagram to show the **pattern** of the magnetic field. **(2 marks)**

    (b) Draw arrows on each diagram to show the **direction** of the magnetic field. **(2 marks)**

2 The magnetic field around a solenoid is similar in shape to the magnetic field of which of these magnets?

    ☐ **A** ball magnet   ☐ **C** circular magnet

    ☐ **B** bar magnet   ☐ **D** horseshoe magnet **(1 mark)**

3 (a) Give **two** factors that affect the strength of the magnetic field around a current-carrying wire.

      > **Guided**

      The strength of the magnetic field depends on the size of the current in the wire

      and the .... **(2 marks)**

    (b) The graphs below show how the strength of a magnetic field varies with two variables. Label the x-axes on the diagrams below with either 'Current' or 'Distance from the wire' to show how they vary with the strength of a magnetic field. **(2 marks)**

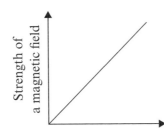 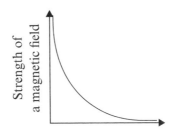

4 In an experiment a student measures the magnetic field strength B at a distance of 15 cm from a wire carrying a current of 1.2 A. The experiment is repeated at distances of 7.5 cm and 30 cm.

    (a) State whether the strength of the magnetic field increases or decreases at:

      (i) 7.5 cm from the wire ....

      (ii) 30 cm from the wire. .... **(2 marks)**

    (b) The student then changes the distance from the wire to 15 cm to take further measurements of the magnetic field strength but changes the current for the first reading to 0.6 A and for the second reading to 2.4 A. State whether the strength of the magnetic field increases or decreases at:

      (i) 0.6 A ....

      (ii) 2.4 A .... **(2 marks)**

Extended response – Magnetism and the motor effect

Describe how an experiment could show the effect and strength of a magnetic field around a long straight conductor and what would be observed when the circuit was connected.

> You will be more successful in extended response questions if you plan your answer before you start writing.
>
> The question asks you to give a detailed explanation of the magnetic field generated by a current-carrying conductor. Think about:
>
> - how you would safely connect the conductor to enable circuit measurements to be taken
> - the methods you could use to determine the direction of a magnetic field
> - the shape of the magnetic field that you would expect to find
> - how you would interpret the field patterns of current-carrying conductors
> - the variable that would influence the strength of the magnetic field around a current-carrying conductor
> - how the influence of the magnetic field of a current-carrying conductor changes.
>
> You should try to use the information given in the question.

....

....

....

....

....

....

....

....

....

....

....

....

....

....

....

....

....

....

.... **(6 marks)**

Transformers

1 Two types of transformers are used in the National Grid. Name them and describe their use.

A step-up transformer ....

....

A step-down transformer ....

.... **(2 marks)**

2 The National Grid transmits electricity from power stations at 400 000 volts (400 kV).

(a) Explain why this voltage is used to transmit
electricity over long distances.

> Remember that increasing the voltage decreases the current.

....

.... **(3 marks)**

(b) State **one** hazard of transmitting electricity at 400 000 V.

.... **(1 mark)**

3 Which link between a part of the National Grid and its function is correct?

Part of National Grid	Function
step-down transformer	transmits electrical energy
National Grid system	decreases voltage
power station	increases voltage
step-up transformer	generates electrical energy

.... **(1 mark)**

4 A step-down transformer with 100% efficiency has a
potential difference of 300 V across the primary coil
and a current of 0.5 A. The secondary coil carries a
current of 10 A. Calculate the potential difference
across the secondary coil. Give the unit.

> You may find this equation useful: $V_P \times I_P = V_S \times I_S$.
> You will need to rearrange it.

potential difference = ... unit.................. **(4 marks)**

Extended response – Electromagnetic induction

Describe how electrical energy is transferred from a power station to a home.

> You will be more successful in extended response questions if you plan your answer before you start writing.
>
> The question asks you to describe the transmission of electrical energy. Think about:
>
> - why electrical energy is transferred at high voltages from power stations
> - where and why step-up transformers are used
> - why high voltages are used in the transmission of electricity
> - where and why step-down transformers are used
> - the assumption made about transformers in the transmission of electricity.
>
> You should try to use the information given in the question.

....

....

....

....

....

....

....

....

....

....

....

....

....

....

....

....

....

....

.... **(6 marks)**

Changes of state

1   Draw a line from each property to the correct state of matter, and from the state of matter to the correct intermolecular forces. Two lines have been drawn for you.

Property	State	Intermolecular forces
Particles move around each other.	solid	some intermolecular forces
Particles cannot move freely.	liquid	almost no intermolecular forces
Particles move randomly.	gas	strong intermolecular forces

(2 marks)

2   (a)   Describe a feature that the three states of matter have in common.

.... **(1 mark)**

(b)   Describe **two** significant differences between the states of matter.

> Different states of matter exist due to differences in energy stores which affects the way the particles behave.

....

.... **(2 marks)**

3   Which statement describes the energy changes that take place when ice melts and then refreezes?

☐   A   Energy is transferred to surroundings → further energy is transferred to surroundings.

☐   B   Energy is transferred to the ice → energy is transferred to surroundings.

☐   C   Energy is transferred to surroundings → energy is transferred to the ice.

☐   D   Energy is transferred to the ice → energy remains in the system.   **(1 mark)**

4   Explain why the temperature stops rising temporarily when a liquid is heated to its boiling point and heating continues.

> Guided

At boiling point the liquid changes state so the energy applied after boiling point
is reached ....

....

.... **(3 marks)**

5   Water is put into the freezer and turns to ice at 0 °C. Describe what happens, in terms of the energy stored, as the temperature continues to fall to −18 °C.

> Remember that the energy stores change as temperature falls.

....

....

....

.... **(3 marks)**

Density

1   A timber centre has to calculate the density of large pieces of wood.

    (a) Calculate the density of a large block of pine wood with a mass of 1650 kg and a volume of 3 m³.

> Recall the equation that links density, mass and volume.

density of wood block = ... kg/m³   (2 marks)

    (b) Calculate the density of a block of elm wood that has a mass of 4 kg and a volume of 0.005 m³.

density of wood block = ... kg/m³   (2 marks)

2   Select the **two** correct statements for density.

    ☐ **A** Density is constant for a substance at constant temperature.

    ☐ **B** Density is related to how the atoms are arranged in a substance.

    ☐ **C** Density changes with increased mass of a substance.

    ☐ **D** Density is calculated using force and volume.   **(2 marks)**

3   A metal block measuring 0.1 m × 0.25 m × 0.15 m has a density of 3000 kg/m³. Calculate the mass of the block. Give your answer in kilograms.

mass of metal block = ... kg   (4 marks)

4   Marco says that all liquids must have lower densities than solids because liquid particles have more kinetic energy and so liquids take up more volume per unit mass than solids. Ella disagrees and says that because solid icebergs float on liquid water, she thinks that Marco must be wrong. Discuss the scientific approaches of these students.

> **Guided**

Marco has approached this problem by ....

....

....

Ella has approached this problem by ....

....

Both students should ....

....   (5 marks)

Investigating density

1 When determining the density of a substance you need to measure the volume of the sample.

   (a) State which other quantity you need to measure.

   .... **(1 mark)**

   (b) Give an example of how you could measure this quantity.

   .... **(1 mark)**

2 The volume of a solid object may be determined by two methods.

   (a) Describe both methods.

   ....

   ....

   ....

   .... **(2 marks)**

   (b) Explain why one method may be preferable to the other.

   ....

   ....

   ....

   .... **(2 marks)**

3 (a) Describe the method that can be used to find the density of a liquid.

   Guided   Place a measuring cylinder on a ....

   ....

   Add the ....

   Record the mass of the ....

   .... **(3 marks)**

   (b) Describe the technique to read the volume of the liquid accurately.

   ....

   .... **(2 marks)**

   (c) Calculate the density of a liquid with a mass of 121 g and a volume of 205 cm^3. Give the unit.

> You may find this equation useful:
> density = mass ÷ volume   ($\rho = m \div V$)

                                   density = ... unit .... **(2 marks)**

Energy and changes of state

1   State what is meant by the term specific latent heat.

....

.... **(1 mark)**

2   Calculate how much energy is required to heat 800 g of water from 30 °C to 80 °C. Take the specific heat capacity of water to be 4200 J/kg °C.

> You may find this equation useful: $\Delta Q = m \times c \times \Delta\theta$ (where $\Delta\theta$ = change in temperature).

energy required = ... J   **(3 marks)**

3   Calculate the amount of energy needed to melt 35 kg of ice. Take the specific latent heat of fusion of water to be 336 000 J/kg.

> You may find this equation useful: $Q = m \times L$

energy required = ... J   **(2 marks)**

4   (a)   Add the following labels to the graph in the boxes provided: melting, evaporating, solid, gas.   **(2 marks)**

> Guided

(b)   Explain what is happening at these stages to result in no rise in temperature.

> Consider the bonds between particles.

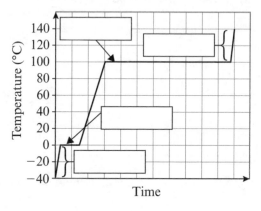

The energy being transferred to the material is breaking .... and, as a result, the

material undergoes a .... **(2 marks)**

5   Calculate the amount of energy needed to change 0.5 kg of water at 20 °C to steam at 100 °C. The latent energy of vaporisation for water is 2 265 000 J/kg.

> Calculate the amount of energy needed to heat the water from 20 °C to 100 °C, to turn it into steam at 100 °C, and then add the two together.

energy needed = ... J   **(5 marks)**

Thermal properties of water

1   Water is widely used in cooling systems because of its relatively high specific heat capacity compared with some other liquids.

    (a) State the definition of the term specific heat capacity.

    .... **(1 mark)**

    (b) Give the equation for specific heat capacity.

    .... **(1 mark)**

2   (a)   Describe an experiment that could be set up to measure the specific heat capacity of water using an electric water heater, a beaker and a thermometer.

> **Guided**

> Remember 'pre-experiment' steps, e.g. zero the balance to eliminate the mass of apparatus before measuring substances, take a starting temperature reading before heating and decide on the range or type of measurements to be taken.

    Place a beaker on a balance, ...

    Take a start reading of .... Place the electrical

    heater .... Take a temperature

    reading .... until the water...................................... **(5 marks)**

    (b)   Suggest how you can determine the amount of thermal energy supplied to the heater by the electric current.

    ....

    ....

    .... **(2 marks)**

    (c)   Explain how this experiment could be improved to give more accurate results.

    ....

    ....

    .... **(2 marks)**

3   A known mass of ice is heated until it becomes steam. The temperature is recorded every minute. Describe how to use the data to identify when there are changes of state.

    ....

    ....

    .... **(2 marks)**

4   Identify **two** hazards and subsequent safety measures that are common to both experiments to determine specific heat capacity and specific latent heat.

    ....

    ....

    .... **(2 marks)**

Pressure and temperature

1   Explain what is meant by temperature.

> Consider the movement of particles.

.... **(1 mark)**

2   (a)   Complete the table below showing some equivalent values in kelvin and degrees Celsius.

Kelvin (K)	Degrees Celsius (°C)
	0
	−18
373	

(3 marks)

   (b)   (i)   Use the particle model to explain what would happen if a substance were to reach 0K and what this temperature is known as.

....

....

.... **(3 marks)**

      (ii)   Give the value of the Celsius scale at which absolute zero occurs.

absolute zero = ... °C   **(1 mark)**

3   In an experiment a fixed-volume container of 100 g of helium gas is warmed from −10°C to 30°C.

> Guided

   (a)   Describe what happens to the velocity of the helium particles as a result of increasing temperature.

As the temperature increases the particles will move faster because ....

.... **(2 marks)**

   (b)   Explain how this affects the pressure on the container walls.

....

....

.... **(2 marks)**

   (c)   State what happens to the average kinetic energy of the particles as the temperature increases.

.... **(1 mark)**

4   In a fixed volume of air the temperature in kelvin is increased by a factor of four. Describe how this affects the average kinetic energy of the air particles.

....

....

.... **(2 marks)**

Extended response – Particle model

The transfer of thermal energy store from a building may be reduced through the use of thermal insulation. Use the particle model to explain how insulation affects the transfer of thermal energy store in various parts of a building.

> You will be more successful in extended response questions if you plan your answer before you start writing.
>
> The question asks you to give a detailed explanation of thermal energy transferred out of a building and ways of reducing this. Think about:
>
> - important areas in a building from where thermal energy is transferred
> - the methods of thermal energy transfer through solids, fluids and a vacuum
> - types of insulators and how they reduce the transfer of thermal energy
> - examples of where to use insulators to reduce the transfer of thermal energy from a building.

....

....

....

....

....

....

....

....

....

....

....

....

....

....

....

....

....

....

....

.... **(6 marks)**

Elastic and inelastic distortion

1 Draw a line from each force pair to the correct distortion it produces.

Force pair
push forces (towards each other)
pull forces (away from each other)
clockwise and anticlockwise

Distortion
stretching
bending
compression

(3 marks)

2 Give an example where each of the following may occur:

(a) tension

.... (1 mark)

(b) compression

.... (1 mark)

(c) elastic distortion

.... (1 mark)

(d) inelastic distortion.

.... (1 mark)

3 A student investigates loading two aluminium beams each with an elastic limit at 50 N. Beam 1 is tested to 45 N. Beam 2 is tested to 60 N. Predict what you would expect the beams to look like after the experiment. Explain your answer.

> Guided

mass — aluminium beam

After testing, beam 1 would ....

....

Beam 2 would ....

.... (4 marks)

4 Car manufacturers use crumple zones to make cars safer. Suggest how crumple zones work in a crash.

> Think about how energy is absorbed to protect the passengers in the event of a crash.

....

....

....

....

.... (3 marks)

Springs

1   State what is meant by the term elastic when describing an object experiencing a force.

....

....

.... **(2 marks)**

> **Guided**

2   A spring is stretched from 0.03 m to 0.07 m, within its elastic limit. Calculate the force needed to stretch the spring. State the unit. Take the spring constant to be 80 N/m.

extension = 0.07 m −

force = ... *× extension =* ....

force = ... *unit* .... **(3 marks)**

3   Deduce the spring constant that produces an extension of 0.04 m when a mass of 2 kg is suspended from a spring. Take *g* to be 10 N/kg.

> Be careful not to confuse mass with force.

☐   **A**   0.02 N/m

☐   **B**   0.08 N/m

☐   **C**   50 N/m

☐   **D**   500 N/m   **(1 mark)**

4   (a)   Calculate the force needed to stretch a spring by 15 cm. The spring has a spring constant of 200 N/m.

> Remember to change the unit of the extension of the spring.

force = ... N   **(2 marks)**

  (b)   Calculate the energy transferred to the spring in (a).

> You may find this equation useful:
> energy = ½ × spring constant × extension²

energy transferred = ... J   **(2 marks)**

Forces and springs

1 (a) Describe how to set up an experiment to investigate the elastic potential energy stored in a spring using a spring, a ruler, masses or weights, a clamp and a stand.

> **Practical skills** Include a step to make sure the spring is not damaged during the experiment.

....

....

....

....

....

.... **(4 marks)**

(b) Explain why it is important to check that the spring is **not** damaged during the experiment.

....

....

.... **(2 marks)**

Guided (c) Explain how the data collected must be processed before a graph can be plotted. Assume masses are used and measurements are made in mm.

Masses must be converted to force (N) by using $W = m \times g$ or $F = m \times g$

The extension of the spring must be ....

....

Extension measurements should be .... **(3 marks)**

(d) Describe how a graph plotted from this experiment can be used to calculate:

   (i) the elastic potential energy stored in the spring

   .... **(1 mark)**

   (ii) the spring constant k.

   .... **(1 mark)**

(e) Write the equation to calculate the energy stored by the spring.

   .... **(1 mark)**

2 Explain the difference between the length of a spring and the extension of a spring.

....

.... **(1 mark)**

Extended response – Forces and matter

1 Forces can change the shape of an object. Explain how forces can result in the two types of distortion and give examples of where distortion may be useful.

> You will be more successful in extended response questions if you plan your answer before you start writing.
>
> The question asks you to give a detailed explanation of the how forces acting on an object can cause it to change shape. Think about:
>
> - how forces cause an object to change shape
>
> - elastic and inelastic distortion
>
> - how energy may be stored and recovered through change of shape
>
> - the result of energy not being recovered through change of shape
>
> - the condition under which metal springs exhibit elastic distortion
>
> - the relationship between force and extension in the elastic behaviour of springs.

....

....

....

....

....

....

....

....

....

....

....

....

....

....

....

....

.... **(6 marks)**

Answers

Biology

1. Plant and animal cells

1 (a) B (1)

   (b) C (1)

2 (a) carry out respiration (1), releasing energy for cell processes (1)

   (b) Mitochondria release energy and all cells need energy (1), but only leaf (and stem) cells are exposed to light and so have chloroplasts for photosynthesis (1).

3 Cell membrane controls what enters and leaves the cell (1); cell wall helps to support the cell / helps it keep its shape (1).

4 Ribosomes are where proteins are made (1); pancreatic cells produce large amounts of proteins but fat cells do not (1).

2. Different kinds of cell

1 C (1)

2 (a) A = acrosome (1); B = flagellum (1)

   (b) A contains enzymes to digest a way through the egg cell membrane (1); B is used to move the bacterium towards a food source (1).

3 Epithelial cells line tubes (such as trachea) (1). Mucus traps dirt / dust / bacteria (1) and cilia move mucus along the tubes away from the lungs (1).

3. Microscopes and magnification

1 Light microscopes magnify less than electron microscopes (1). The level of cell detail seen with an electron microscope is greater (1), because electron microscopes have a higher resolution (1).

2 (a) because it has a nucleus (1) and eukaryotic cells have nuclei (1)

   (b) (i) $(23 / 5) \times 2$ (1) = 9.2 µm (1)

       (ii) $(4 / 5) \times 2$ (1) = 1.6 µm (1)

   (c) Nuclei are large enough to be seen with a light microscope (1) but mitochondria are too small and can be seen only with an electron microscope (1) because it has a higher resolution / greater magnification (1).

3 (a) light microscope: 2.5 µm × 1000 (1) = 2500 µm (or 2.5 mm) (1); electron microscope: 2.5 µm × 100 000 = 250 000 µm (or 250 mm or 25 cm or 0.25 m) (1) (Note that you get the mark for correct use of the formula just once even though you use it twice.)

   (b) The electron microscope (1) because it would show more detail / has the correct resolution (1).

4. Dealing with numbers

1 picometre, nanometre, micrometre, millimetre, metre (1)

2 5 picometres (1), 0.25 grams (1), 0.00025 kilograms (1), 2500 millimetres (1)

3 true (1), false (1), false (1), true (1)

4 (a) $(30.9/1\,000\,000) \times 1\,000\,000$ (1) = 30.9 nm (1)

   (b) $(163/250\,000) \times 1\,000\,000$ (1) = 652 nm (1)

   (c) $(7.8/800) \times 1000$ (1) = 9.75 µm (1)

5. Using a light microscope

1 (a) (i) to reflect light through the slide (1)

       (ii) to hold the slide in place (1)

       (iii) to move the objective up and down a long way (1)

   (b) (i) because it could crash into the slide (1)

       (ii) because it could permanently damage eyesight (1)

   (c) (i) a desk / bench / built-in lamp (1)

       (ii) Two from: always start with the lowest power objective under the eyepiece (1); clip the slide securely on the stage (1); move the slide so the cell you need is in the middle of the (low-power) view (1); use only the fine focusing wheel with the high-power objective (1)

2 Three from: go back to using the low-power objective (1); find the part you need and bring it back to the centre view (1); focus on it with the coarse focus (1); return to the high-power objective (1) and use the fine focus wheel to bring the part into focus (1).

6. Drawing labelled diagrams

1 (a) Three from: the drawing is in pen rather than pencil (1); the title is incomplete (1); the magnification is not given (1); label lines are not drawn with a ruler (1) and cross each other (1); not enough cells are shown (1) and they are not drawn to scale (1); shading should not be used (1); lines have been crossed out rather than rubbed out (1) and are ragged rather than clear (1); the cell membrane can't be seen with the light microscope (1)

   (b) Clear drawing of all / most of the cells (1); cells not of interest drawn just as outlines (1); detail of representative sample of cells (1); and avoidance of mistakes from 1 (a) (1)

2 width of image = 45 mm (1) so magnification = 45 / 0.113 (1) = ×398 (or 400) (1)

7. Enzymes

1 The shape of the active site of invertase matches the shape of sucrose but not lactose (1), so invertase cannot combine with lactose and catalyse its digestion (1).

2 (a) Answer in the range 43–45 °C (1)

   (b) D (1)

   (c) As the temperature increases, the rate of collisions increases (1) between the substrate and active site (1).

3 Optimum pH of pepsin is about 2 (1); optimum pH of trypsin is about 8 (1); these are the same as the pH in the stomach and small intestine (1).

8. pH and enzyme activity

1 (a)

pH	2	4	6	8	10
Time (min)	> 10	7.5	3.6	1.2	8.3
Rate (min)	0	0.13	**0.28**	**0.83**	**0.12**

(**2 marks** for all 5 correct, **1 mark** for 3 correct)

(b)

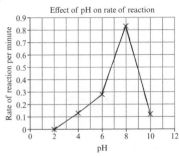

correctly plotted points (1), points joined by straight lines / line of best fit (can be smooth curve) (1)

(c) Two from: use a water bath to control temperature (1); repeat several times and take a mean (1); use a more accurate method to determine if the film is clear (1); use more intermediate pH values (1)

9. The importance of enzymes

1

Enzyme	Digests	Product(s)
amylase	starch	sugars / maltose
lipase	**lipids**	**fatty acids and glycerol**
protease	**proteins**	amino acids

1 mark for each correct row.

2 (a) Many different enzymes are needed because they are specific for different food molecules (1); digestion breaks down the food molecules into molecules small enough to be absorbed (1).

   (b) Synthesis reactions occur too slowly (1); enzymes are biological catalysts and speed up reactions (1).

3 (a) protease **(1)**

(b) The enzyme is denatured / active site destroyed **(1)** at higher temperatures **(1)**, so it would not digest stains as well / would be less active **(1)**.

4 Both involve enzymes **(1)**; digestion involves breaking down large molecules to form small molecules **(1)** but synthesis involves producing large molecules from smaller molecules **(1)**.

10. Getting in and out of cells

1 movement of particles **(1)** from high concentration to low concentration / down a concentration gradient **(1)**

2 One mark for each correct row to **4 marks**:

Feature	Diffusion	Active transport
Involves the movement of particles	✓	✓
Requires energy		✓
Can happen across a partially permeable membrane	✓	✓
Net movement down a concentration gradient	✓	

3 (a) Osmosis is the net movement of water molecules **(1)** across a partially permeable membrane **(1)** from a low solute concentration **(1)** to a high solute concentration **(1)**.

(b) Diffusion **(1)**; because movement is from a high concentration to a low concentration / down a concentration gradient **(1)**

(c) Glucose must be moved against a concentration gradient **(1)** by active transport that requires energy **(1)**.

11. Osmosis in potatoes

1 Four from: Cut pieces of potato, making sure size / length is the same **(1)**; measure mass **(1)**; leave in solution for 20 minutes / same time **(1)**. Remove from the solution, then measure mass again **(1)**. Blot dry before each weighing **(1)**.

2 (a) missing change in mass: 0.25 **(1)**; missing percentage change = (–0.15/2.58) × 100 = –5.8 % **(1)**

(b) points plotted ± half square **(1)**; line of best fit **(1)**

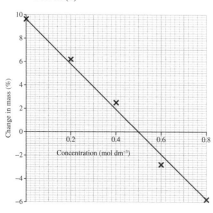

(c) answer in the range 0.45 mol dm⁻³ to 0.55 mol dm⁻³ **(1)**

12. Extended response – Key concepts

Answer could include the following points: **(6)**
Similarities: both have:
- cell membrane, which controls what enters and leaves the cell
- cell wall, which strengthens the cell
- cytoplasm, in which cell reactions happen
- ribosomes, where protein synthesis happens.

Differences:
- plant cells have mitochondria, where respiration happens, but bacterial cells do not
- plant cells have chloroplasts, where photosynthesis happens, but bacterial cells do not
- DNA, which contains genetic information, is contained in a nucleus in plant cells, but is found in a single chromosome / loop and in plasmids in bacterial cells
- cell wall in plant cells is made from cellulose but from different substances in bacterial cells
- plant cells have a permanent vacuole filled with cell sap, which helps to keep the cell rigid, but bacterial cells do not.

13. Mitosis

1 (a) B **(1)**

(b) interphase, prophase, metaphase, anaphase, telophase **(1)**

2 to produce new individuals by asexual reproduction **(1)**; for growth **(1)**; for repair **(1)**

3 (a) A = anaphase **(1)**; B = metaphase **(1)**

(b) A is because chromatids are being pulled to each pole **(1)**; B is because chromosomes are lined up along the middle of the cell **(1)**

14. Cell growth and differentiation

1 (a) zygote **(1)**

(b) mitosis **(1)**

2 (a) meristem / root tip / shoot tip **(1)**

(b) Vacuoles take in water by osmosis **(1)** and this causes the cell to elongate **(1)**.

3 (a)

Type of specialised cell	Animal or plant
sperm	animal
xylem	plant
ciliated cell	animal
root hair cell	plant
egg cell	animal

3 marks for 4 or 5 correct, **2 marks** for 3 correct, **1 mark** for 2 correct.

(b) Plants: mesophyll cell / guard cell / phloem **(1)**. Animals: small intestine cell / hepatocyte / red blood cell / nerve cell / bone cell / (smooth) muscle cell **(1)**.

4 (a) Cells become specialised **(1)** to perform a particular function. **(1)**

(b) Many different kinds of specialised cells **(1)** can carry out different processes more effectively **(1)**.

15. Growth and percentile charts

1 (a) C **(1)**

(b) 47.5 – 46.0 **(1)** = 1.5 cm (± 0.2 cm) **(1)**

2 (a) 15.35 – 12.75 = 2.60 g **(1)**; (2.60 / 12.75) × 100 = 20.4% **(1)**

(b) Any suitable, such as: height **(1)**, measured with a ruler ensuring the stem is vertical **(1)**; shoots / leaves **(1)** by counting number **(1)**

16. Stem cells

1 (a) A **(1)**

(b) (i) meristem **(1)**

(ii) tips of root **(1)** and tips of shoot **(1)**

2 (a) to replace damaged / worn-out cells **(1)**

(b) Differentiated cells cannot divide / embryonic stem cells can divide, to produce other kinds of cell **(1)**.

3 (a) Embryonic stem cells could be stimulated to produce nerve cells **(1)** then transplanted into the patient's brain. **(1)**

(b) (i) Does not destroy embryos / patient's immune system will not reject them **(1)**

(ii) May cause cancer / may not differentiate into nerve cells **(1)**

17. Neurones

1 **One mark** for one correct; **two marks** for all three correct. Motor neurone – carries impulses from the central nervous system to effectors; relay neurone – carries impulses from one part of the central nervous system to another; sensory neurone – carries impulses to the central nervous system

2 A, axon endings; B, axon; C, cell body; D, dendron; E, myelin sheath; F, receptor cells (in skin) (all correct, **3 marks**; 4 or 5 correct, **2 marks**; 2 or 3 correct, **1 mark**)

3 *The axon is long so it* can carry impulses over long distances **(1)**. The axon has a myelin sheath which is an electrical insulator / prevents impulses passing to neighbouring neurones **(1)**. *The nerve ending transmits impulses to* effectors/glands/muscles **(1)**.

4 (a) Myelin sheath speeds up transmission **(1)** because the impulse jumps from one gap to another **(1)**.

(b) Their movement would be impaired / made difficult **(1)** because the nerve impulses to muscles would be slower / can move between adjacent neurones **(1)**.

18. Responding to stimuli

1 In any order: innate **(1)**; automatic **(1)**; rapid **(1)**

2 (a) synapse **(1)**

(b) neurone Y **(1)**; because it is carrying impulses to an effector / muscle **(1)**

(c) When an electrical impulse reaches the end of neurone X it causes the release of neurotransmitter **(1)** into the gap between the neurones. This substance diffuses **(1)** across the synapse / gap **(1)** and causes neurone Y to generate an electrical impulse **(1)**.

3 (a) Three from: Stimulus is detected by receptors **(1)**; a nerve impulse travels along a sensory neurone **(1)** then through

a relay neurone in the brain / CNS / spinal cord (1) and along a motor neurone to an effector (1).

(b) light / movement (1) because it causes the eyelid to blink (1)

19. Extended response – Cells and control

Answer could include the following points: (6)

- Stages of mitosis described as part of the cell cycle.
- Production of genetically identical daughter cells.
- Diploid number maintained in all cells except gametes.
- Involves replication of DNA.
- Description of cell differentiation.
- Examples of specialised cell types.
- Importance of stem cells: in embryo to produce all different kinds of cell in the body; in adult for growth and repair.

20. Meiosis

1   (a) (i) half the number of chromosomes / one set of chromosomes (1)

   (ii) sex cells (1)

   (b) male: sperm (1); female: egg / ovum (1)

2   (a) 10 (1)

   (b) Each daughter cell has only half of chromosomes / genes / DNA from each parent (1) so different daughter cells have different combinations of chromosomes / genes / DNA (1).

3   (a) DNA replication (1)

   (b)

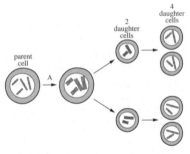

2 daughter cells, then 4 daughter cells (1), one of each pair consisting of duplicated chromosomes in 2 daughter cells (1); 4 daughter cells contain one copy of each pair (1)

4   Mitosis maintains the diploid number (1) and produces cells that are identical to the parent cell (1). It is used for growth (1). Meiosis creates gametes that have half the number of chromosomes (1). Fertilisation restores the diploid number (1).

21. DNA

1   (a) genome (1)

   (b) A chromosome consists of a long molecule of DNA packed with proteins (1); a gene is a section of DNA molecule / section of chromosome that codes for a specific protein (1). DNA is the molecule containing genetic information that forms part of the chromosomes (1).

2   (a) double helix (1)

   (b) (i) 4 (1)

   (ii) weak hydrogen bonds between complementary bases (1)

3   (a) The structure consists of repeated nucleotides / monomers (1).

   (b) A base (1), B sugar / ribose (1), C phosphate (1)

4   A (1)

22. Genetic terms

1   1 mark for each correct answer. (4)

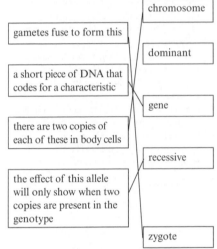

	chromosome
gametes fuse to form this	
	dominant
a short piece of DNA that codes for a characteristic	
	gene
there are two copies of each of these in body cells	
	recessive
the effect of this allele will only show when two copies are present in the genotype	
	zygote

2   (a) (i) different forms of the same gene that produce different variations of the characteristic (1)

   (ii) Genotype shows the alleles that are present in the individual, e.g. Bb or BB (1), whereas phenotype means the characteristics that are produced, e.g. brown eyes or blue eyes (1).

   (b) bb (1), BB (1), Bb (1)

   (c) bb (1) because to have blue eyes she must have two recessive alleles (1)

23. Monohybrid inheritance

1   (a) correct gametes (1); correct genotypes (1)

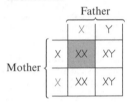

	T	t
T	TT	Tt
t	Tt	tt

   (b) 25% of the offspring from this cross will be short (1). I know this because tt is short (1) and one in four of the possible offspring is tt. (1)

   (c) $\frac{3}{4}$ or 75% (1)

2   (a) Completed Punnett square:

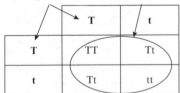

		Parent genotype Gg	
	Parent gametes	G	g
Parent genotype gg	g	Gg	gg
	g	Gg	gg

Parent gametes correct (1); offspring genotypes correct (1)

   (b) 20 (1)

24. Family pedigrees

1   C (1)

2   (a) two (1)

   (b) one (1)

   (c) Person 4 does not have cystic fibrosis. This means that she must have one dominant allele from her father (1). But she must have inherited a recessive allele from her mother (1). This means that her genotype is Ff (1).

   (d) Two healthy parents (person 3 & person 4) (1) produce a child (person 8) with CF (1).

25. Sex determination

1   (a) X (1); the girl has two X chromosomes, one from each parent (1)

   (b) (i) 1 mark for parental sex chromosomes and 1 mark for all possible children's chromosomes

		Father	
		X	Y
Mother	X	XX	XY
	X	XX	XY

   (ii) female (1)

2   (a) 50% / ½ / 0.5 (1); depends on which sperm fertilises the egg (1) as half the sperm will carry a male sex chromosome / Y chromosome and half the sperm will carry a female sex chromosome / X chromosome (1).

   (b) The statement is not correct (1); the probability of having a child who is a boy is always 50% (1).

26. Variation and mutation

1   (a) Students in a year 7 class will show differences in mass caused by genetic variation (1) as well as environmental variation. (1) Can be opposite order.

   (b) Identical twins will show differences caused only by environmental variation (1).

2   (a) mean height = (181 + 184 + 178 + 190 + 193 + 179) / 6 (1) = 184.2 cm (1)

   (b) Four from: height is determined partly by genetic factors (1) and partly by environmental factors (1) such as nutrition (1); different children have inherited different alleles from their parents (1); parents have different heights so will pass on different alleles for height (1); the mean height of the children is greater than that of the parents because of better nutrition / they take after their father more than their mother (1)

3   (a) a change in an organism's DNA (1) such as a change in a gene / sequence of bases (1)

(b) Two from: no effect (**1**), small effect (**1**), significant effect (**1**) on phenotype

27. The Human Genome Project

1 (a) the sequence of bases on all human chromosomes (**1**)

(b) Advantages, two from: a person at risk from a genetic condition will be alerted (**1**); distinguishing between different forms of disease (**1**); tailoring treatments for some diseases to the individual (**1**)

Disadvantages: people at risk of some diseases may have to pay more for life insurance (**1**); it may not be helpful to tell someone they are at risk from an incurable disease (**1**)

2 Advantages, any two from: she could have earlier / more frequent screening for breast cancer (**1**), she could consider surgery to remove the breast / mastectomy (**1**), her doctor might prescribe drugs to reduce the risk of developing cancer (**1**); disadvantages, any two from: it might make her more worried / anxious (**1**), just because she has the mutation doesn't mean she will develop breast cancer (**1**), could have unnecessary surgery / medication (**1**)

28. Extended response – Genetics

Answer could include the following points: (**6**)

- Mutation is a change in the DNA sequence.
- Most mutations do not affect the phenotype / some mutations have a small effect on the phenotype.
- A single mutation can (rarely) significantly affect the phenotype.
- Human Genome Project maps the DNA / base pairs in the human genome.
- Human genome project helps to identify the gene for p53.
- This can be linked to the risk of developing cancer.
- Different mutations might increase the risk by different amounts.
- So the risk / whether they carry a harmful allele for a particular individual can be estimated.

29. Evolution

1 (a) (i) Different individuals have different characteristics (because they have different genes/alleles) (**1**).

(ii) changes in conditions such as change in availability of food, shelter, change in climate, new predator, new disease (**1**)

(b) because variation in some individuals makes them better at coping with change (**1**) and so more likely to survive (**1**)

2 Within a population of a species there is variation. (**1**) Those members of the species that are most adapted will survive / those that are less well adapted die (**1**).

3 It will help to classify the new species (**1**) and to find out which other organisms the new species is related to (**1**).

4 There is variation in the amount of antibiotic resistance in a population of bacteria (**1**); the most resistant take the longest to die (**1**), so stopping early means the most resistant will survive and reproduce (**1**) so that all the new population of bacteria will be resistant (**1**).

30. Human evolution

1 Three from: toe arrangement (**1**), length of arms (**1**), brain size (**1**), skull shape (**1**)

2 (a) Two from: the older the species, the smaller its brain volume (**1**); negative correlation / as years before present became less, brain volume increased OR positive correlation – as time 'increases' brain volume increases (**1**); greatest increase in brain volume between 2.4 and 1.8 million years ago (**1**); increase in brain volume not linear, but increased by 500 cm^3 in 2.6 million years (**1**)

(b) an increase in brain volume / size (**1**) to at least 550 cm^3 (**1**)

3 (a) The ages of the rock layers where the tool was found can be dated (**1**) by measuring the amount of radiation in the layers (**1**).

(b) Three from: smooth area in palm of hand (**1**), will not cut / damage hand (**1**); chipped section away from hand (**1**) (as it) has sharp edges (**1**) for cutting / unlike smooth area (**1**)

31. Classification

1 mouse and goat (**1**) because they share the least recent common ancestor (**1**)

2 Plants are autotrophic feeders but animals are heterotrophic feeders (**1**). Plant cells have cells walls but animal cells do not (**1**). *Could also say* animal cells do not contain chlorophyll / plant cells contain chlorophyll (**1**)

3 Panther / *Panthera pardus* and wolf / *Canis lupus* (**1**); because they both belong to the same (kingdom, phylum, class and) order (**1**)

4 (a) C (**1**)

(b) genetic research / research on genes (**1**)

32. Selective breeding

1 (a) Plants or animals with certain desirable characteristics are chosen to breed together (**1**) so that their offspring will inherit these characteristics (**1**).

(b) Pigs with lower body fat are crossed (**1**); offspring with low body fat are selected and crossed (**1**); repeated for many generations until a lean breed is produced (**1**).

2 (a) high yield so can feed more people (**1**); low fertiliser requirement so no need to apply fertiliser / reduce cost (**1**); pest resistant (or example given) so less pest damage / do not need to apply pesticide (**1**).

(b) drought resistant to cope with times of water shortage without dying (**1**); tolerant of high temperature (**1**)

3 Three from: alleles that might be useful in the future might no longer be available (**1**); a new disease might affect all organisms (**1**); selectively bred organisms might not adapt to changes in climate (**1**); animal welfare might be harmed (**1**)

33. Genetic engineering

1 (a) D (**1**)

(b) (i) One from: resistance to herbicides (**1**); production of beta-carotene (**1**)

(ii) One from: increased crop yields (**1**); less insecticide needed (**1**)

(iii) One from: may kill insect species that are not pests (**1**); less food for birds (**1**); gene for insect resistance may transfer to another species of plant (**1**)

2 The gene from a jellyfish (**1**) is cut out using enzymes. (**1**) This gene is transferred to a mouse (**1**) embryo cell, and inserted into a chromosome. The embryo is then allowed to develop as normal.

3 Four from: GM bacteria produce human insulin not pig insulin (**1**); so this will be more effective/is the right form of insulin/is less likely to cause adverse reactions (**1**); can be produced in large quantities by the bacteria (**1**); this means that it can be produced at low cost (**1**); some people would have ethical/ religious objections to having insulin from pigs (**1**)

34. Extended response – Genetic modification

Answer could include the following points: (**6**)

- Selective breeding produces plants with desirable features.
- Genetic engineering produces plants with desirable features.
- Selective breeding takes many generations but genetic engineering is much quicker.
- Selective breeding has happened for many years but genetic engineering is a recent process.
- Desirable features include resistance to drought/pesticides/herbicides/insects.
- Tissue culture is used to produce clones of a plant.
- Tissue culture does not change plants (unlike selective breeding and genetic engineering).
- Tissue culture allows many plants with desirable features to be produced.
- Some people have ethical objections to genetic engineering / GM plants.
- Selective breeding is widely accepted / people usually do not have ethical objections to it.

35. Health and disease

1 (a) being free from disease and eating and sleeping well (**1**)

(b) how you feel about yourself (**1**)

(c) how well you get on with other people (**1**)

2 (a) Communicable: ✓influenza, ✓tuberculosis; ✓*Chlamydia*; Non-communicable: ✓lung cancer, ✓coronary heart disease

(**3 marks** for 5 correct, **2 marks** for 3 or 4 correct, **1 mark** for 1 or 2 correct)

(b) Communicable: rapid variation in number of cases over time / cases localised (**1**); non-communicable: number of cases change gradually / cases more widespread (**1**)

3 (HIV) causes damage to the immune system (**1**); reduced immune response / immunity (**1**)

4 (a) Three from: a virus infects a body cell (**1**) and takes over the body cell's DNA (**1**) causing the cell to make toxins (**1**) or damages the cell when new viruses are released (**1**).

(b) Bacteria can release toxins (**1**) and can invade and destroy body cells (**1**).

36. Common infections

1 (a) Zimbabwe (1); 15.1–14.3 = 0.8% decrease (1)

 (b) All countries show a decrease in the percentage of 15 to 49 year olds with HIV (1); one example of such a trend is: all percentages have dropped somewhere between 0.3 and 2.9% (1).

2 (a) D (1)

 (b) Two from: leaf loss (1); bark damage (1); dieback of top of tree (1)

3

Disease	Type of pathogen	Signs of infection
cholera	**bacterium**	watery faeces
tuberculosis	bacterium	persistent cough – may cough up blood
malaria	**protist**	**fever, weakness, chills and sweating**
HIV	**virus**	mild flu-like symptoms at first

(all correct for **3 marks**, 3 correct for **2 marks**, 2 or 1 correct for **1 mark**)

4 (a) bacterium (1)

 (b) Two from: inflammation in stomach (1); bleeding in stomach (1); stomach pain (1)

37. How pathogens spread

1 C (1)

2

Disease	Pathogen	Ways to reduce or prevent its spread
Ebola haemorrhagic fever	**virus (1)**	Keep infected people isolated; wear full protective clothing while working with infected people or dead bodies.
tuberculosis	bacterium	**Ventilate buildings to reduce chance of breathing in bacteria / diagnose promptly and give antibiotics to kill bacteria / isolate infected people (1).**

3 Boil water before drinking / wash hands after using toilet (1) because bacteria are spread in water / by touch (1).

4 (a) The bacteria are spread in water (1); in developed countries water is treated to kill pathogens / good hygiene prevents their spread (1).

 (b) to prevent being infected by the Ebola virus (1) because Ebola virus is present in body fluids of infected people even after death (1)

38. STIs

1 an infection spread by sexual activity (1)

2 B (1)

3

Mechanism of transmission	Precautions to reduce or prevent STI
unprotected sex with an infected partner	using condoms during sexual intercourse (1)
sharing needles with an infected person (1)	supplying intravenous drug abusers with sterile needles
infection from blood products	**screening blood transfusions (1)**

4 (a) Screening helps identify an infection (1) so people can be treated for the disease / take extra precautions to prevent transmission (1).

 (b) HIV is a virus (1) and viruses cannot be treated with antibiotics (1).

39. Human defences

1 (a) Skin acts as a physical barrier that stops microorganisms getting into the body (1).

 (b) Hydrochloric acid in the stomach kills pathogens (1).

 (c) (i) lysozyme (1)

  (ii) kills bacteria (1) by digesting their cell walls (1)

2 (a) (i) mucus (1)

  (ii) sticky so traps bacteria / pathogens (1)

 (b) (i) cilia (1)

  (ii) The cilia on the surface of these cells move in a wave-like motion (1) and this moves mucus and trapped pathogens out of lungs (1) towards the back of the throat where it is swallowed (1).

 (c) Mucus travels down to into the lungs carrying pathogens (1) because the cilia cannot move and take the pathogens back up to the throat (1).

40. The immune system

1 lymphocytes (1)

2 Pathogens have substances called antigens (1) on their surface. White blood cells called lymphocytes (1) are activated if they have antibodies (1) that fit these substances. These cells then divide many times to produce clones / identical cells (1). They produce large amounts of antibodies that stick to the antigens / destroy the pathogen (1).

3 (a) Lymphocytes producing antibodies against measles virus are activated (1); these lymphocytes divide many times (1), so concentration of antibodies increases (1) then decreases when the viruses have all been destroyed (1).

 (b) Some of the lymphocytes stay in the blood as memory lymphocytes (1); these respond / divide after infection (1), so the number of lymphocytes producing the antibodies against the measles virus increases rapidly (1).

 (c) (i) (The girl had not been exposed to the chickenpox virus before because) line B is similar in size and shape to line A (1), which was for a first infection with measles / the line would be higher if it was a second infection (1).

  (ii) The concentration of antibodies increased faster / to a higher concentration (1), so the measles viruses were destroyed before they could cause illness / symptoms / disease (1).

41. Immunisation

1 (a) artificial immunity to a pathogen (1) by using a vaccine (1)

 (b) A vaccine contains antigens from a pathogen (1), often in the form of dead / weakened pathogens (1).

 (c) The vaccine produces memory lymphocytes (1) so, if the person is exposed to the disease, the memory lymphocytes produce a very rapid secondary response (1), so it is very unlikely they will become ill (1).

2 (a) 2003

 (b) The number of cases would increase (1) because fewer babies were immunised (1).

42. Treating infections

1 (a) C (1)

 (b) Antibiotics kill bacteria / inhibit their cell processes (1) but do not affect human cells (1).

2 The pharmacist's advice would be not to take the penicillin (1). The man's cold is due to a virus, so the penicillin will not be effective in combating the infection (1).

3 (a) Sinusitis is (probably) not caused by a bacterial infection (1).

 (b) Same number of patients got better (after 14 days) without antibiotics (1), although the patients taking antibiotics may have got better (a little) more quickly (1).

43. New medicines

1 (a) 3, 1, 5, 2, 4 (all correct = **2 marks**, 4 correct = **1 mark**)

 (b) (i) testing in cells or tissues to see if the medicine can enter cells and have the desired effect (1); testing on animals to see how it works in a whole body / has no harmful side effects (1)

  (ii) by testing in a small number of healthy people (1)

 (c) Medicine is tested on people with the disease that it will be used to treat (1) so that the correct dose can be worked out (1) and to check for side effects in different people (1).

2 (a) Large number of subjects make the data valid (1) and repeatable (1); OR side effects will be seen only in small numbers (1) so it is easier to notice with a large trial group (1); OR there are different stages of the trial (1) and each step needs a different group of people (1).

(b) The medicine appears to be effective in nearly 400 people with high blood pressure (1); this reduction is much greater than those in the placebo group (1). You could also say: the medicine seems to have very little adverse effect on the blood pressure of those in the 'normal' group (so it is effective).

44. Non-communicable diseases

1 An infectious disease is caused by a pathogen (1) and is passed from one person to another (1). A non-communicable disease is not passed from one person to another (1).

2 Three from: inherited / genetic factors (1); age (1); sex (1); ethnic group (1); lifestyle (e.g. diet, exercise, alcohol, smoking) (1); environmental factors (1)

3 (a) (i) Bangladeshi men (1)

(ii) black women (1)

(b) Four from: the prevalence of CHD increases with age (1); overall the prevalence is higher in men than in women (1), but prevalence is similar in black men and women (1); Bangladeshi men have the highest prevalence but Bangladeshi women are in the middle (1); ethnic group seems to be a bigger factor in men than in women (1); the prevalence in all ethnic groups is very similar in the 40–49 age group (1)

45. Alcohol and smoking

1 (a) Ethanol is a drug that is toxic / poisonous to cells (1). It is broken down by the liver and harms liver cells (1). Too much alcohol over a long period causes cirrhosis / liver disease (1).

(b) because it is caused by how we choose to live (1)

2 Two from: because carbon monoxide in cigarette smoke (1) reduces how much oxygen the blood can carry to the baby (1), leading to low birth weight in babies / other abnormalities (1)

3 (a) Two from: cardiovascular disease (1); lung cancer (1); respiratory / lung disease (1)

(b) Substances in cigarettes cause blood vessels to narrow (1) which increases the blood pressure (1) leading to cardiovascular disease (1).

46. Malnutrition and obesity

1 (a) too little of one or some nutrients in the diet (1)

(b) Four from: Anaemia increases with increasing age (1) in both men and women (1), but whereas there is an increase in females from 1–16 and 17–49 (1) followed by a decline (1), in males the lowest age groups are 17–49 and 50–64 (1).

2

Subject	Weight (kg)	Height (m)	BMI
person A	80	1.80	24.7
person B	90	1.65	33.1
person C	95	2.00	23.8

All 3 correct = 2 marks, 2 correct = 1 mark

3 (a) A = 0.975 ÷ 1.02 = 0.96 (1), B = 0.914 ÷ 1.06 = 0.86 (1), 1 mark for 2 decimal places for both.

(b) Man A (1) because his waist-to-hip ratio is greater than 0.90 (1)

47. Cardiovascular disease

1 (a) Two from: lifestyle changes (1); medication (1); surgery (1)

(b) Two from: give up smoking (1); take more exercise (1); eat a healthier diet (lower fat, sugar and salt) (1); lose weight (1)

(c) because cardiovascular disease reduces life expectancy (1) and can be fatal before treatment can be given (1)

2 Lifestyle changes – Benefits: may reduce chances of getting other health conditions / the cheapest option. Drawbacks: may not work effectively.

Medication – Benefits: starts working immediately / cheaper and less risky than surgery. Drawbacks: needs to be taken long term / may not work well with other medication the person is taking.

Surgery – Benefits: usually a long-term solution. Drawbacks: there is a risk the person will not recover after the operation / expensive / more difficult to do than giving medication.

(3 marks for 6 correct, 2 marks for 4–5 correct, 1 mark for 2–3 correct)

3 Surgery can help prevent heart attacks / strokes (1), but costs more than inserting a stent (1) and surgery has more risk (e.g. risk of infection) (1). However, it can be a long-term solution / other suitable conclusion (1).

48. Extended response – Health and disease

Answer could include the following points: (6)
- Communicable diseases caused by infection with a pathogen.
- Non-communicable diseases are not caused by infection.
- Non-communicable diseases may have genetic causes.
- Non-communicable diseases may have environmental / lifestyle factors, e.g. diet / nutrition / alcohol / smoking.
- But (poor) diet / nutrition can also increase risk of catching infections.
- Treatment of communicable diseases largely through medicines, e.g. antibiotics.
- Non-communicable diseases can be treated with medicines but also through lifestyle changes.

49. Photosynthesis

1 Plants or algae are photosynthetic organisms / producers (1) so they are the main producers of biomass (1) and animals have to eat plants / algae (1).

2 (a) carbon dioxide + water (1) → glucose + oxygen (1)

(b) The product of photosynthesis / glucose has more energy than the reactants (1) because energy is transferred from the surroundings / light (1).

3 (a) Light is required for photosynthesis (1) because only parts of the leaf exposed to light produced starch (1).

(b) Chlorophyll / chloroplasts are required for photosynthesis (1) because only green areas / areas with chlorophyll or chloroplasts produced starch (1).

50. Limiting factors

1 This is a factor or variable that stops the rate of something increasing/changing (1). The rate will only increase if this factor is increased (1).

2 (a) temperature (1)

(b) Add algal balls to hydrogen carbonate solution (1). Leave in light for a set amount of time, e.g. 2 hours (1). Compare the colour change against standard colours (1).

3 (a) Increasing the carbon dioxide concentration increases the rate of photosynthesis (1).

(b) Adding carbon dioxide means that it will no longer be a limiting factor (1), so the plants can make more sugars needed for growth (1).

(c) You could the increase the temperature (1) as this would make photosynthesis happen faster / more quickly (1).

51. Light intensity

1 (a) points plotted accurately (1); curve of best fit drawn (1)

(b) 76 (± 2) (1)

(c) As the distance increases, the rate of bubbling decreases (1), not a linear relationship (1).

(d) (i) Take care not to touch the bulb if it is hot (1).

(ii) Place a water tank next to the bulb if it is hot (1) to help prevent heat from the lamp reaching the test tube (1). OR Use a ruler to make sure that the lamp is at the measured distance (1) because differences in distance will change the light intensity (1).

(e) You could use the light meter to measure the light intensity (1) at each distance and then plot a graph of rate of bubbling against light intensity (1).

52. Specialised plant cells

1 (a) phloem (1)

(b) A There are holes (1) to let liquids flow from one cell to the next (1).

B There is a small amount of cytoplasm (1) so there is more space for the central channel (1).

(c) Mitochondria supply energy (1) for active transport (of sucrose) (1).

2 (a) xylem (1)

(b) Three from: the walls are strengthened with lignin rings to prevent them from collapsing (1); no cytoplasm means that there is more space for water (1); pits in the walls allow water and mineral ions to move out (1); no end walls means that they form a long tube so water flows easily (1)

53. Transpiration

1 (a) Transpiration is the loss of water (1) by evaporation from the leaf surface (1).

(b) stomata (in the leaf) (1)

(c) (i) moves faster (1) because a faster rate of water loss from leaves (1)

(ii) moves more slowly (1); stomata covered so a lower rate of water loss (1)

2 (a) Guard cells take in water by osmosis (1) so they swell, causing the stoma to open (1); when the guard cells lose water they become flaccid / lose rigidity and the stoma closes (1).

(b) The stomata are open during the day, so water is lost by transpiration (1) faster than it can be absorbed by the roots (1). Water is lost from the vacuoles and the plant wilts. At night, the stomata close so water is replaced (1).

54. Translocation

1 (a) the movement of sucrose around a plant (1)

(b) A (1)

2 (a) Radioactive carbon dioxide is supplied to the leaf of a plant (1). The radioactive carbon / sucrose will then be detected in the phloem (1) and eventually incorporated into starch in the potato (1).

(b) Translocation will stop / sucrose will not be transported from the leaf / radioactivity will not be detected elsewhere in the plant (1) because translocation uses active transport (1).

3

Structure or mechanism	Transport of water	Transport of sucrose
xylem	X	
phloem		X
pulled by evaporation from the leaf	X	
requires energy		X
transported up and down the plant		X

(1 mark for each correct row.)

55. Water uptake in plants

1 One mark for each correct row:

Information	Increased light intensity	Increased temperature
stomata become more open	✓	
stomata become more closed		
water molecules have less energy		
water molecules have more energy		✓
rate of evaporation increased	✓	✓
rate of evaporation decreased		

2 (a) The rate of evaporation was higher when the fan was on (1), because the movement of air removes water more quickly from the leaves (1), increasing the concentration gradient from leaf to air (1).

(b) The rate of evaporation became quicker than the rate at which the plant could take up water (1); the stomata of the plant closed (1) to prevent evaporation from occurring / conserve water (1).

(c) The volume of the tube is calculated using $\pi r^2 l$. volume = $(\pi \times 0.25^2 \times 90)$ = 17.67 mm^3 (1)
rate = 17.67 / 5 = 3.5 mm^3 / min (1)

56. Extended response – Plant structures and functions

Answer could include the following points: (6)

● Guard cells can take in water by osmosis, and swell.

● This happens in the light.

● When the guard cells swell, they become rigid and the stomata / stoma opens.

● Water is lost from the leaf through open stomata.

● This helps move water and dissolved mineral ions up the stem by transpiration.

● At night, the guard cells lose water and the stomata / stoma closes.

● Closing stomata when it is dark and no photosynthesis is occurring helps reduce water loss.

57. Hormones

1 (a) Hormones are produced by endocrine glands (1) and are released into the blood (1). They travel round the body until they reach their target organ (1), which responds by releasing another chemical substance (1).

(b) Hormones have long-lived effects; nerves have short-term effects (1). Nerve impulses act quickly; hormones take longer (1).

2 A = hypothalamus (1), B = pituitary (1), C = thyroid (1), D = pancreas (1), E = adrenal (1), F = testis (1) and G = ovary (1)

3

Hormone	Produced in	Site of action
FSH and LH	pituitary gland	ovaries
insulin and glucagon	pancreas	liver, muscle and adipose (fatty) tissue
adrenalin	adrenal gland	various organs, e.g. heart, liver, skin
progesterone	ovaries	uterus
testosterone	testes	male reproductive organs

1 mark for each correct row

58. The menstrual cycle

1 Two from: oestrogen (1); progesterone (1); FSH (1); LH (1)

2 (a) A = menstruation (1); B = ovulation (1)

(b) any time between day 14 and about day 17 (1)

(c) The lining of the uterus breaks down (1) and is lost in a bleed or period (1).

3 (a) Pills, implants or injections release hormones that prevent ovulation (1), and thicken mucus at the cervix (1), preventing sperm (1) from passing.

(b) (i) These figures are maximum values (1) when the methods are used correctly (1).

(ii) All three ticked for 'Reduces chance of pregnancy'; only male condom ticked for 'Protects against STIs' (1 mark for each correct row)

59. Blood glucose regulation

1 3, 5, 1, 4, 2. (All 5 correct = 3 marks, 3 correct = 2 marks, 1 correct = 1 mark.)

2 (a) $((7.0 - 4.6) \div 4.6) \times 100$ (1) = 52% (1)

(b) Liver / muscle cells take up glucose (1) because high blood glucose causes pancreas to release insulin (1).

(c) Glucose is being taken up by muscles more rapidly (1) because they use more glucose during exercise (1).

60. Diabetes

1 (a) As the BMI increases the percentage of people with diabetes increases (1) so there is a positive correlation (1).

(b) (i) BMI = $88 \div 1.8^2$ = 27.2 (1); he (is overweight so) has an increased risk of Type 2 diabetes (1) but not the highest risk (1).

(ii) W:H ratio = $104 \div 102$ = 1.02 which is obese (1) so he has a high risk of developing Type 2 diabetes (1) because there is a correlation between W:H ratio and risk of Type 2 diabetes (1).

2 (a) Controlling diets will help to control the number of people who are obese (1). Fewer obese people means fewer people with diabetes (1).

(b) (i) In Type 1 diabetes no insulin is produced so has to be replaced with injections (1) but in Type 2 diabetes organs don't respond to insulin (1).

(ii) A large meal means a higher blood glucose concentration (1) so more insulin is needed to reduce the glucose concentration (1).

61. Extended response – Control and coordination

Answer could include the following points: (6)

● Cause of Type 1 diabetes: immune system has damaged insulin-secreting cells in pancreas, so no insulin produced.

● Cause of Type 2 diabetes: insulin-releasing cells may produce less insulin and target organs are resistant / less sensitive to insulin.

● Link risk of Type 2 diabetes with obesity / BMI / waist : hip ratio.

● Treat Type 1 diabetes by injecting insulin. Amount of insulin injected can be changed according to the blood glucose concentration.

● Treat Type 2 diabetes by diet (eating healthily and reduced sugar) and exercise.

● Treat more severe Type 2 diabetes with medicines to reduce the amount of glucose the liver releases or to make target organs more sensitive to insulin.

62. Exchanging materials

1 (a) kidneys / nephrons (1) to maintain constant water level / osmoregulation (1)

(b) kidneys / nephrons (1); urea is a toxic waste product (1)

2 in the lungs (1); oxygen is needed for respiration (1); carbon dioxide is a waste product (1)

3 (a) The surface of the small intestine is covered with villi (1). These help by increasing the surface area available for absorption (1).

(b) This makes the absorption of food molecules more efficient / effective (1) by reducing the distance that the molecules have to diffuse (1).

4 Four from: The flatworm is very flat and thin (1) which means that it has a large surface area : volume ratio (1); the earthworm is cylindrical so has smaller surface area : volume ratio (1); every cell in the flatworm is close to the surface (1); in the earthworm diffusion has to happen over too great a distance / through too many layers of cells (1).

63. Alveoli

1 (a) Oxygen diffuses from the air in alveoli into the blood in capillaries (1). Carbon dioxide diffuses from the blood into the air (1).

(b) Millions of alveoli create a large surface area for the diffusion of gases (1). Each alveolus is closely associated with a capillary (1). Their walls are one cell thick (1). This minimises the diffusion distance (1).

2 maintains concentration gradient (1) which maximises the rate of diffusion (1)

3 Three from: breathlessness / shortness of breath / similar (1); less oxygen in blood than normal (1) so less respiration / energy (1); increased carbon dioxide concentration reduces pH (1) which affects enzyme-controlled reactions (1)

64. Blood

1 One mark for one correct; two marks for two correct; three marks for four correct. Plasma – carries other blood components; platelet – involved in forming blood clots; red blood cell – carries oxygen; white blood cell – part of the body's immune system (1)

2 (a) nucleus (1)

(b) haemoglobin (1)

(c) Their biconcave shape gives them a large surface area (1) for diffusion to happen efficiently. They are also flexible, which lets them fit through narrow blood vessels/capillaries (1).

3 urea (1); carbon dioxide (1)

4 Platelets respond to a wound by triggering the clotting process (1); the clot blocks the wound (1) and prevents pathogens from entering (1).

5 (a) Infections are caused by pathogens (1); lymphocytes produce antibodies (1) that stick to pathogens and destroy them (1).

(b) Phagocytes surround foreign cells (1) and digest them (1).

65. Blood vessels

1 (a) An artery has thick walls (1). These walls are composed of two types of fibres: connective tissue (1) and elastic fibres (1).

(b) Wall stretches as blood pressure rises / heart ventricles contract (1) and recoils (*not contracts!*) when blood pressure falls / heart ventricles relax (1).

2 (a) Thin walls / only one-cell thick (1) run close to almost every cell (1).

(b) faster diffusion of substances (1) because short distance / large surface area (1)

3 (a) (i) Blood flows at low pressure (1) so no need for elastic wall of arteries / need wide tube in veins (1).

(ii) Muscles contract and press on veins (1); blood pushed towards heart because valves prevent flow the wrong way (1).

(b) Veins have a thinner muscle wall than arteries (1) so it is easier to get the needle in (1). OR Veins contain blood under lower pressure (1) so taking blood is more controlled (1).

66. The heart

1

Blood vessel	Carries blood:	
	from	to
aorta	heart	body
pulmonary artery	heart	lungs
pulmonary vein	lungs	heart
vena cava	body	heart

1 mark per correct line

2 (a) because it acts as a pump (1) and muscles contract to pump the blood (1)

(b) order of parts: (vena cava) right atrium, right ventricle, pulmonary artery (lungs), pulmonary vein, left atrium, left ventricle (aorta) (names all correct for 2 marks, 4 correct for 1 mark; additional mark for correct order)

3 (a) right ventricle (1); pumps blood to the lungs / pulmonary artery (1)

(b) heart valve closes when ventricle contracts (1); prevents backflow (1)

(c) has to pump harder (1) to get blood all round body (1), not just to lungs (1)

67. Aerobic respiration

1 (a) oxygen and glucose (2)

(b) Diffusion is the movement of substances from high to low concentration (1).

2 (a) mitochondria (1)

(b) Respiration is an exothermic process (1) and transfers energy by heating (1).

(c) active transport / muscle contraction / other appropriate use (1)

3 (a) glucose + oxygen → carbon dioxide + water (1)

(b) capillaries (1)

4 (a) Respiration releases energy (1) so that metabolic processes that keep the organism alive can continue (1).

(b) Two from: Plants cannot use energy from sunlight directly for metabolic processes (1) so they need energy from respiration for this purpose (1) during the day as well as at night (1).

68. Anaerobic respiration

1 (a) Aerobic respiration releases more energy (1) per molecule of glucose (1).

(b) The body needs energy more quickly than aerobic respiration can supply (1); it cannot get enough oxygen to respiring cells (1).

2 (a) Heart rates will increase gradually / remain low during early laps (1) and increase rapidly during final sprint (1) because energy demand is low at first then increases significantly (1). You could also say that adrenalin might increase heart rate during early laps.

(b) Two from: to keep heart rate relatively high (1) so that lactic acid is removed from muscles (1); because oxygen is needed to release energy needed to get rid of lactic acid (1)

3 (a) Three from: oxygen consumption increases during exercise (1) but reaches a maximum value (1); no more oxygen can be delivered for aerobic respiration (1); increased energy needed comes from anaerobic respiration (1)

(b) During exercise there is an increase in the concentration of lactic acid (1); after exercise, extra oxygen is needed to break down lactic acid (1).

69. Rate of respiration

1 (a) maintains a constant temperature (1); because temperature can affect enzymes / change the rate of reaction (1)

(b) absorbs carbon dioxide produced by the seeds (1) so that this doesn't interfere with the movement of the blob of water (1).

(c) allows the pressure to be released between experiments (1); so the blob of water is pushed back to the start position (1)

2 (a) Movement of the blob of water indicates uptake of oxygen for use in respiration (1), so measuring the movement of the blob at intervals (1) allows the rate of respiration to be calculated by dividing distance moved by time taken (1).

(b) Use the water bath at a range of temperatures (1); measure distance moved by the blob over a particular time (1) and repeat several times at each temperature (1).

70. Changes in heart rate

1 (a) Stroke volume is the volume of blood pumped from the heart in one beat (1).

(b) (i) cardiac output = stroke volume × heart rate = 60 × 75 (1) = 4500 (1) cm^3 / min (1)

(ii) Cardiac output increases (1); then two from: cells need to respire faster / need more oxygen and glucose (1); increased stroke volume / more blood needed for respiring cells (1), so heart rate must increase (1).

239

Answers

2 (a) $100 - 80 = 20$ **(1)**; $20 / 80 \times 100 = 25\%$ **(1)**

(b) highest demand for oxygen / glucose / respiration **(1)**

(c) Rearrange the equation to give stroke volume = cardiac output / heart rate **(1)** = $4000 / 50 = 80$ **(1)** cm³.

71. Extended response – Exchange

Answer could include the following points: **(6)**
Outline of route:
- vena cava → right atrium → right ventricle → pulmonary artery → (capillaries in) lungs → pulmonary vein → left atrium → left ventricle → aorta → rest of the body / capillaries in the body → vena cava

Answer might also include:
- valves in heart / veins prevent backflow of blood
- deoxygenated blood enters / leaves right side
- oxygenated blood enters / leaves left side
- walls of left side of heart are thicker than right side.

72. Ecosystems and abiotic factors

1 **1 mark** for each

Term	Definition
Community	A single living individual
Organism	All the living organisms and the non-living components in an area
Population	All the populations in an area
Ecosystem	All the organisms of the same species in an area

2 (a) south side = 323.5 **(1)**; north side = 227.6 **(1)**

(b) She is correct **(1)** because the mean percentage cover is 39% on the south side compared with 4% on the north side / 10 times greater **(1)**.

(c) temperature **(1)** because it affects enzymes / rate of reactions **(1)** OR humidity **(1)** because water required for photosynthesis / other cell processes **(1)**

73. Biotic factors

1 (a) the living parts of an ecosystem **(1)**

(b) Two from: so that they can become the new alpha male **(1)**; to gain fighting skills **(1)**; to become stronger **(1)**

(c) Food can often be scarce in their habitat **(1)**, so large groups need to split into different areas in order to find enough food **(1)**.

2 The peacock has large, attractive tail feathers **(1)**; it competes with other males for mates **(1)**, so large showy tails are more attractive to female peahens **(1)**. You could also suggest that the large tail feathers can be used to help scare away other male peacocks who may compete for females.

3 (a) The trees emerge through the canopy to get more light **(1)** for more photosynthesis **(1)**.

(b) The trees have deep / extensive roots **(1)** to collect minerals **(1)**.

74. Parasitism and mutualism

1 **1 mark** for each

Statement	Parasitism only	Mutualism only	Both parasitism and mutualism
There is interdepence, where the survival of one species is closely linked with another species.			X
One species lives inside the intestine of another and absorbs nutrients from the digested food.	X		
One species lives inside another and receives food from the host. In return, the host receives nutrients.			X

2 Cleaner fish get food by eating parasites from the skin of sharks **(1)**. This helps the shark because it reduces the risk of the shark being harmed by the parasites **(1)**.

3 Scabies mite lives in the host and causes it harm **(1)**; no benefit to the host **(1)**

75. Fieldwork techniques

1 (a) He could use a 1 m × 1 m quadrat **(1)**, which he could throw at the flower bed to choose a random location **(1)**.

(b) Using the same area means that his experiment is a fair test **(1)**.

(c) He could look at more than one area each day **(1)** and take an average number of slugs **(1)**.

2 (a) Find the total number of plants counted and the number of quadrats **(1)** and calculate a mean number of clover plants **(1)**.

(b) total size of field = $100 \times 65 = 6500$ m² **(1)**, so number of clover plants = 6500×7 **(1)** $= 45\,500$ **(1)**

3 Place quadrats at regular intervals along the transect **(1)** and measure the percentage cover of broad-leaved plants in each quadrat **(1)**. Record a named abiotic factor (light intensity / temperature) at each quadrat position **(1)**.

76. Organisms and their environment

1 (a) Draw a line from the seashore up the beach (at right angles to the sea line) **(1)**; place quadrat at regular intervals along the line **(1)**; count the limpets in each quadrat area **(1)**.

(b) $(10 + 8 + 9) / 3$ **(1)** $= 9$ **(1)**

(c) The number of limpets goes down as you travel further from the sea **(1)**; this decrease is linear with the distance / it drops by 4 limpets for every 0.5 m distance travelled **(1)**; limpets are more likely to survive if they live nearer the seashore **(1)**.

2 (a) Instead of placing a quadrat every 2 m, the scientist could place a quadrat every 0.5 m / 1 m / smaller distance **(1)** and use a smaller **(1)** quadrat than before.

(b) Two from: the number of bluebells increases to a maximum around 8 m into the wood **(1)**; less light is available for photosynthesis **(1)** and fewer nutrients / water available deeper in the wood where there are more trees **(1)**

77. Human effects on ecosystems

1 Advantage – reduces fishing of wild fish **(1)**
Disadvantage – one of: the waste can pollute the local area changing conditions so that some local species die out **(1)**; diseases from the farmed fish (e.g. lice) can spread to wild fish and kill them **(1)**

2 Advantage – may provide food for native species **(1)**; OR may increase biodiversity **(1)**
Disadvantage – one of: may reproduce rapidly as they have no natural predators in the new area **(1)**; may out-compete native species for food or other resources **(1)**

3 (a) $145 - 15 = 130$; $(130 / 15) \times 100$ **(1)** $= 867\%$ **(1)**

(b) increasing population **(1)**; more food / crops needed **(1)**

(c) Excess fertiliser can be leached / washed into rivers / lakes **(1)**, causing eutrophication **(1)**.

78. Biodiversity

1 (a) replanting forests where they have been destroyed **(1)**

(b) Two from: restores habitat for endangered species **(1)**; reduces carbon dioxide concentration in the air (as trees photosynthesise) **(1)**; reduces the effects of soil erosion **(1)**; reduces range of temperature variation **(1)**

2 Some species are valuable to humans **(1)** because they are a source of new drugs / are wild varieties of crop plants / source of genes **(1)**.

3 The numbers of trees will increase because there are fewer deer to eat them **(1)**. This means that there will be more food for birds / bears / rabbits / insects **(1)**. There will be more rabbits because there are fewer coyotes to kill / eat them **(1)**. If there are more rabbits, there will be more food for coyotes / hawks / predators **(1)**. More trees also mean that there will be more habitats for birds **(1)** and less soil erosion **(1)**.

79. The carbon cycle

1 (a) photosynthesis **(1)**

(b) respiration **(1)**

(c) combustion **(1)**

(d) decomposition **(1)**

2 Microorganisms are decomposers **(1)**; they convert complex carbon-containing molecules into carbon dioxide (by respiration), which is released into the atmosphere **(1)**.

3 (a) Fish carry out respiration (**1**); respiration releases carbon dioxide into the water (**1**); plants absorb the carbon dioxide (**1**), which is used in photosynthesis (**1**).

(b) Any three from: if there are not enough fish / snails / aquatic animals in the tank there will not be enough carbon dioxide (**1**), so there is less photosynthesis by plants (**1**), so less oxygen is released for fish / snails / aquatic animals (**1**); less food for animals as fewer plants (**1**); plants and fish / snails / aquatic animals die (**1**)

80. The water cycle

1 (a) Three from: evaporation from land (**1**), sea (**1**) and transpiration from plants (**1**); animal sweat (**1**); animal breath (**1**)

(b) Water vapour condenses to form clouds (**1**); water cools to form precipitation / rain / snow (**1**) that returns the water to Earth (**1**).

2 A lot of water evaporates from golf courses so this will lead to more water in the atmosphere (**1**); water levels fall in the river as water is removed for watering the golf course (**1**) so animals or plants living in the river might die (**1**).

3 Advantage: sea water is made potable / safe to drink (**1**)

Disadvantage: needs a lot of energy / fuel / it is expensive (**1**)

81. The nitrogen cycle

1 C (**1**)

2 (a) nitrogen fixation by soil bacteria (**1**)

(b) Three from: nitrates are absorbed by roots (**1**) by active transport (**1**) because plants need nitrogen for making amino acids / proteins (**1**) but can only take in nitrogen in the form of nitrate / ammonium (ions / salts) (**1**)

(c) Amount of nitrate in soil is reduced (**1**) because bacteria convert nitrates in the soil into nitrogen gas in the air (**1**).

3 (a) Three from: Plants such as clover have nitrogen-fixing bacteria in their roots (**1**) so they can be grown and ploughed back into the soil (**1**); where they are decomposed (**1**) to add nitrates (**1**).

82. Extended response – Ecosystems and material cycles

Answer could include the following points: (**6**)

- Fish farming can reduce biodiversity by introducing just one species.
- Waste and diseases can affect wild populations.
- Introduction of non-native species might lead to competition with native species.
- Reduces fishing of wild fish.
- Fertilisers can cause eutrophication leading to loss of biodiversity in nearby water.
- You could also talk about conservation, reforestation, captive breeding.

Chemistry

83. Formulae

1 C (**1**)

2 first box ticked (**1**), last box ticked (**1**)

3 **1 mark** for each correct formula: water, H_2O; carbon dioxide, CO_2; methane, CH_4; sulfuric acid, H_2SO_4; sodium, Na

4 An element is a substance made from atoms (**1**) with the same number of protons (**1**).

5 (a) 3 (**1**)

(b) 7 (**1**)

6 (a) There is a negative sign in the formula (**1**).

(b) The 3 shows that there are three oxygen atoms (**1**); the 2 shows that the ion has two (negative) charges (**1**).

84. Equations

1 B (**1**)

2 table completed to show: copper carbonate – reactant (**1**); copper oxide *and* carbon dioxide – product (**1**)

3 sodium hydroxide + hydrochloric acid → sodium chloride + water (**1**)

4 **1 mark** for all four state symbols in the correct order: (s), (l), (aq), (g)

5 **1 mark** for each correctly balanced equation:

(a) $2Cu + O_2 \rightarrow 2CuO$ (**1**)

(b) $2Al + Fe_2O_3 \rightarrow Al_2O_3 + 2Fe$ (**1**)

(c) $Mg + 2HNO_3 \rightarrow Mg(NO_3)_2 + H_2$ (**1**)

(d) $Na_2CO_3 + 2HCl \rightarrow 2NaCl + H_2O + CO_2$ (**1**)

(e) $Cl_2 + 2NaBr \rightarrow 2NaCl + Br_2$ (**1**)

(f) $4Fe + 3O_2 \rightarrow 2Fe_2O_3$ (**1**)

85. Hazards, risk and precautions

1 **1 mark** for each correct new line (if more than four new lines drawn, subtract 1 mark for each extra line):

Symbol	Description
	flammable may easily catch fire
	oxidising agent may cause other substances to catch fire, or make a fire worse
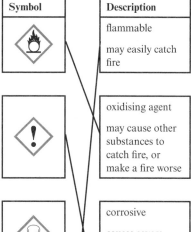	corrosive causes severe damage to skin and eyes
	harmful or irritant health hazard
	toxic may cause death by inhalation, ingestion or skin contact

2 to indicate the dangers associated with the contents (**1**); to inform people about safe-working precautions with these substances (**1**)

3 A hazard is something that could cause damage / harm to someone / something (**1**) or cause adverse health effects (**1**).

4 Risk is the chance that someone or something will be harmed (**1**) if exposed to a hazard (**1**).

5 (a) suitable precaution, e.g. wear gloves / work in a fume cupboard (**1**)

(b) reason to match the precaution in (a), e.g. to avoid skin contact because nitric acid is corrosive / to avoid breathing in nitrogen dioxide (which is toxic) (**1**)

86. Atomic structure

1 D (**1**)

2 A (**1**)

3 **1 mark** for each correct row:

	Protons	Neutrons	Electrons
Nucleus	✓	✓	
Shells			✓

4 2 or 3 correct (**1**), 4 correct (**2**)

Particle	Proton	Neutron	Electron
Relative mass	1	1	1/1836
Relative charge	+1	0	−1

5 Protons and electrons have equal but opposite charges / the relative charge of a proton is +1 and the relative charge of an electron is −1 (**1**); these charges cancel out / add up to zero (**1**).

6 These particles were not discovered until later (**1**).

87. Isotopes

1 The mass number of an atom is the total number of protons and neutrons (in the nucleus) (**1**).

2 A (**1**)

3 An element consists of atoms that have the same number of protons (**1**) in the nucleus, and this is different for different elements (**1**).

4 (a) **1 mark** for each correct row:

Isotope	Protons	Neutrons	Electrons
hydrogen-1	1	0	1
hydrogen-2	1	1	1
hydrogen-3	1	2	1

(b) Isotopes of an element have atoms with the same number of protons (**1**), but different numbers of neutrons (**1**).

5 Some elements have different isotopes (**1**) so their relative atomic masses are a (weighted) mean value (**1**).

Answers

88. Mendeleev's table

1. (a) D **(1)**
   (b) the properties of the elements and their compounds **(1)**

2. (a) one of the following for **1 mark**: elements are in groups; elements are in periods; elements with similar properties are in the same groups
   (b) (i) two of the following for **1 mark each**: Mendeleev's table had fewer elements; did not include the noble gases; was arranged in order of increasing relative atomic mass (not atomic number); did not have a block of transition metals; silver / copper in Group 1; had two elements in some spaces
   (ii) He could predict the properties of undiscovered elements **(1)**.

3. Tellurium has a greater / higher relative atomic mass than iodine does **(1)**.
   However, iodine atoms have more protons than tellurium atoms do **(1)**.

89. The periodic table

1. B **(1)**
2. (a) Group **(1)**
   (b) A and B **(1)**
   (c) A, B, C, D (all four for **1 mark**)
   (d) B and E **(1)**

3. (a) the position of an element on the periodic table **(1)**
   (b) the number of protons **(1)** in an atom's nucleus **(1)**

90. Electronic configurations

1. (a) 2.1 **(1)**
   (b) There are three electrons, so there must be three protons **(1)**, so the four shaded circles must be neutrons **(1)**.
   (c) two shells **(1)**; two electrons in first shell, six in the second **(1)**, e.g.

2. (a) Both have 7 electrons **(1)** in their outer shell **(1)**.
   (b) Fluorine has two occupied shells **(1)** but chlorine has three **(1)**.

3. (a) 2.8.5 **(1)**
   (b) 2.8.8.2 **(1)**

4. Group 0 **(1)** because it has a full outer shell **(1)**

91. Ions

1. D **(1)**
2. (a) $12 - 2 = 10$ **(1)**
   (b) 2.8.8 **(1)**

3. **1 mark** for each correct row:

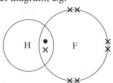

Ion	Atomic number	Mass number	Protons	Neutrons	Electrons
N^{3-}	7	15	7	8	10
K^+	19	39	**19**	**21**	**18**
Ca^{2+}	20	40	**20**	**20**	**18**
S^{2-}	16	32	**16**	**16**	**18**
Br^-	35	80	**35**	**46**	**36**

4. three shells and 2.8.7 electrons in the chlorine atom **(1)**; three shells and 2.8.8 electrons in the chloride ion **(1)**; brackets with negative sign **(1)**, e.g.

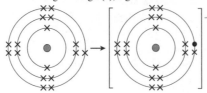

92. Formulae of ionic compounds

1. D **(1)**
2. **1 mark** for each correct formula:

	Cl^-	S^{2-}	OH^-	NO_3^-	SO_4^{2-}
K^+	KCl	K_2S	KOH	KNO_3	K_2SO_4
Ca^{2+}	$CaCl_2$	CaS	$Ca(OH)_2$	$Ca(NO_3)_2$	$CaSO_4$
Fe^{3+}	$FeCl_3$	Fe_2S_3	$Fe(OH)_3$	$Fe(NO_3)_3$	$Fe_2(SO_4)_3$
NH_4^+	NH_4Cl	$(NH_4)_2S$	NH_4OH	NH_4NO_3	$(NH_4)_2SO_4$

3. (a) $2Mg + O_2 \rightarrow 2MgO$ **(1)**
   (b) (i) The nitrogen atom has five electrons in its outer shell **(1)**; it gains three electrons to complete its outer shell / form an ion **(1)**.
   (ii) Mg_3N_2 **(1)**
   (iii) A nitride ion contains only nitrogen **(1)** but a nitrate ion also contains oxygen **(1)**.

4. **1 mark** for each correct name:

	S^{2-}	SO_4^{2-}	Cl^-	ClO_3^-
Name	sulfide	sulfate	chloride	chlorate

93. Properties of ionic compounds

1. A **(1)**
2. (a) + and – signs drawn as shown **(1)**

   (b) There are strong electrostatic forces of attraction **(1)** between oppositely charged ions **(1)**.

3. A lot of heat / energy is needed to break / overcome **(1)** the many / strong ionic bonds / bonds between ions **(1)**.

4. (a) liquid *and* dissolved in water are ticked only **(1)**
   (b) must be able to move around **(1)**

94. Covalent bonds

1. C **(1)**
2. A covalent bond forms when a pair of electrons is shared **(1)** between two atoms **(1)**.
3. (a) There is one covalent bond **(1)** between a hydrogen atom and a fluorine atom in a molecule **(1)**.
   (b) correct diagram, e.g.

   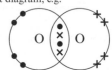

   one pair of dots and crosses in shared area **(1)**; (three) pairs of dots / crosses on F only **(1)**

4. correct diagram, e.g.

   two pairs of dots and crosses in shared area **(1)**; two pairs of dots on one atom and two pairs of crosses on the other atom **(1)**

95. Simple molecular substances

1. C **(1)**
2. (a) C **(1)**
   (b) two from the following, for **1 mark each**: It has a low melting point / lowest melting point; it does not conduct electricity when solid or liquid; it is (almost) insoluble in water.

3. (a) Sulfur hexafluoride molecules are not charged **(1)** and have no electrons that are free to move **(1)**.
   (b) The intermolecular forces between water and sulfur hexafluoride molecules **(1)** are weaker than those between water molecules **(1)** and those between sulfur hexafluoride molecules **(1)**.

96. Giant molecular substances

1. B **(1)**
2. (a) carbon **(1)**
   (b) four **(1)**
   (c) giant molecular / giant covalent **(1)**

3. (a) The layers in graphite can slide over each other **(1)** because there are weak forces between the layers **(1)**.
   (b) Atoms in graphite can only form three covalent bonds **(1)** so graphite has electrons that are delocalised / free to move **(1)**.
   (c) Diamond has a regular / lattice **(1)** structure, and its atoms are joined by many strong bonds / covalent bonds **(1)**.

97. Other large molecules

1 C **(1)**

2 (a) carbon **(1)**

(b) covalent **(1)**

3 (a) Its molecules contain 60 atoms / few atoms **(1)** but giant covalent substances contain very many atoms **(1)**.

(b) Buckminsterfullerene has a simple molecular **(1)** structure so it has weak intermolecular forces / forces between molecules **(1)** that are easily overcome.

4 Graphene has covalent **(1)** bonds in a giant / lattice structure **(1)** and these bonds are strong / need a large amount of energy to break **(1)**.

98. Metals

1 A **(1)**

2 **1 mark** for each correct tick (deduct **1 mark** for each tick over four ticks)

	Low melting points	High melting points	Good conductors of electricity	Poor conductors of electricity
Metals		✓	✓	
Non-metals	✓			✓

3 (a) Fizzing is caused by bubbles of hydrogen gas / hydrogen is a gas **(1)**.

(b) Sodium hydroxide dissolves in water **(1)** so the sodium gradually becomes smaller (as the reaction carries on) **(1)**.

4 (a) (i) + / 2+ inside the circles **(1)**

(ii) delocalised electrons / sea of electrons shown between the circles **(1)**

(b) It has layers of atoms / (positive) ions **(1)** which can slide over each other **(1)**.

(c) Delocalised electrons / free electrons **(1)** can move through the structure / metal **(1)**.

99. Limitations of models

1 B **(1)**

2 (a) A, B, C **(1)**

(b) A, B **(1)**

(c) C, D **(1)**

(d) B **(1)**

(e) C, D **(1)**

3 (a) Unlike a dot-and-cross diagram, a ball-and-stick model shows the shape of the molecule / can be modelled in three dimensions (e.g. using a modelling kit) **(1)**.

(b) Unlike a dot-and-cross diagram, a ball-and-stick model does not show (two of the following, for **1 mark** each): the symbols of the elements; the bonding electrons; the non-bonding electrons.

100. Relative formula mass

1 (a) 71 **(1)**

(b) 18 **(1)**

(c) 64 **(1)**

(d) 102 **(1)**

(e) 53.5 **(1)**

(f) 111 **(1)**

(g) 133.5 **(1)**

2 (a) 74 **(1)**

(b) 78 **(1)**

(c) 164 **(1)**

(d) 132 **(1)**

(e) 342 **(1)**

101. Empirical formulae

1 (a) (i) to make sure that the reaction had finished / all the magnesium had reacted **(1)**

(ii) to let air / oxygen in **(1)**

(b) Use tongs / allow the crucible to cool down **(1)** to prevent skin burns **(1)**.

2 mass of oxygen reacted
= 20.65 g − 20.49 g = 0.16 g **(1)**

Mg	O
0.24/24 = 0.010	0.16/16 = 0.010 **(1)**
0.010/0.010 = 1.0	0.010/0.010 = 1.0 **(1)**

Empirical formula is MgO **(1)**

3 M_r of NO_2 = 14 + (2 × 16) = 46 **(1)**

factor needed = 92 / 46 = 2

Molecular formula is N_2O_4 **(1)**

102. Conservation of mass

1 (a) closed system, because no substances can enter or leave **(1)**

(b) (i) It stays the same **(1)**.

(ii) Mass is conserved in chemical reactions / atoms are not created or destroyed in chemical reactions **(1)**.

2 8.2 − 5.3 = 2.9 g **(1)**

3 234 g **(1)**

4 M_r of O_2 = 32 and M_r of MgO = 40 **(1)**

(1 × 32) = 32 g of O_2 makes (2 × 40) = 80 g of MgO **(1)**

12.6 g of O_2 makes 80 × (12.6/32) g of MgO = 31.5 g of MgO **(1)**

103. Concentration of solution

1 D **(1)**

2 (a) volume = 2500/1000 = 2.5 dm^3 **(1)**

(b) 0.5 dm^3 **(1)**

(c) 0.025 dm^3 **(1)**

3 (a) 25 g dm^{-3} **(1)**

(b) 36.5 g dm^{-3} **(1)**

(c) 5 g dm^{-3} **(1)**

4 concentration =
(10/250) × 1000 = 40 g dm^{-3} **(1)**

5 16 g dm^{-3} **(1)**

6 (a) mixture of a solute and water / solution in which the solvent is water **(1)**

(b) 100 g **(1)**

104. Extended response – Types of substance

The answer may include some of the following points: **(6)**

Graphite uses and properties:

• lubricant because it is slippery / the softest in the table / 10 times softer than copper

• electrodes because it is a good conductor of electricity / conductivity is 100 times less than copper / 100 million times better than diamond.

Diamond uses and properties:

• cutting tools because it is very hard / 100 times harder than copper / 1000 times harder than graphite.

Graphite bonding and structure:

• giant covalent / giant molecular structure
• strong covalent bonds
• each carbon atom is bonded to three other carbon atoms
• layers of carbon atoms
• weak forces between layers
• layers can slide past each other (making it slippery so that it can be used as a lubricant)
• one free electron per carbon atom
• delocalised electrons
• electrons can move through the structure (allowing it to conduct electricity for use as an electrode).

Diamond bonding and structure:

• giant covalent / giant molecular structure
• strong covalent bonds
• each carbon atom is bonded to four other carbon atoms
• three-dimensional lattice structure
• a lot of energy is needed to break the many strong bonds
• rigid structure.

105. States of matter

1 C **(1)**

2 (a) freezing / solidifying **(1)**

(b) condensing / liquefying **(1)**

3 The chemical composition is unchanged **(1)**.

4 **1 mark** for each correct row in the table, to **3 marks** maximum

State of matter	Particles are:			
	Close together	Far apart	Randomly arranged	Regularly arranged
solid	✓			✓
liquid	✓		✓	
gas		✓	✓	

5 (a) gas **(1)**

(b) gas **(1)** because the particles are moving freely / moving fastest / have the most kinetic energy **(1)**

6 liquid **(1)**

7 The arrangement changes from random to regular **(1)** and the movement changes from moving around each other (in groups) to vibrating about fixed positions **(1)**.

106. Pure substances and mixtures

1 D **(1)**

2 The orange juice contains different substances **(1)** mixed together / not chemically joined together **(1)** but a pure substance in the scientific sense contains only one substance / element / compound **(1)**.

Answers

3 (a) The atoms of an element all have the same atomic number / number of protons **(1)** but atoms of Na and Cl_2 have different atomic numbers / numbers of protons **(1)**.

(b) substance containing two or more elements **(1)** *chemically* combined / joined together **(1)**

4 (a) **B** 0.24 and **C** 0.03 **(1)**

(b) None of the samples is pure **(1)**; all contain some residue / dissolved solid / solid mixed in **(1)**.

5 Pure substances (e.g. tin and silver) have a sharp melting point **(1)** but mixtures (e.g. lead-free solder) melt over a range of temperatures **(1)**.

107. Distillation

1 C **(1)**

2 (a) (i) condenser **(1)**

(ii) It decreases / goes down **(1)**.

(b) The temperature of the water increases **(1)** because it is heated up by the vapour / energy is transferred from the vapour to the water by heating **(1)**.

3 (a) ethanol because it has a lower boiling point than water **(1)**; it boils / evaporates first **(1)**

(b) one of the following for **1 mark**:
- More energy is transferred by heating.
- Hot vapour and cold water flow in opposite directions.
- The condenser will be full of cold water / will not contain any air.

108. Filtration and crystallisation

1 second box ticked **(1)**; fourth box ticked **(1)**

Deduct 1 mark for each extra tick above two ticks.

2 (a) $2KI(aq) + Pb(NO_3)_2(aq) \rightarrow 2KNO_3(aq) + PbI_2(s)$

1 mark for correct balancing; **1 mark** for correct state symbols

(b) (i) Its particles are too large to pass through **(1)**.

(ii) to remove excess potassium iodide solution / lead nitrate solution / potassium nitrate solution **(1)**

3 (a) filtration / filtering **(1)**

(b) (i) The *water* evaporates **(1)**; solution becomes saturated / crystals form as more *water* evaporates **(1)**.

(ii) to dry the crystals **(1)**

(c) Heat the solution slowly / do not evaporate all of the water / leave to cool / stop heating before crystals start to form **(1)**.

109. Paper chromatography

1 (a) Pencil does not dissolve in the solvent **(1)**.

(b) mixture because it contains more than one substance / four substances **(1)**; pure substances contain only one substance **(1)**

(c) **A** and **B** **(1)**

(d) The orange squash does contain X because it contains spots with the same R_f values / that move the same distances **(1)** as the two spots in X **(1)**.

(e) It is insoluble in the solvent / it is insoluble in the mobile phase / has very strong bonds with the stationary phase / has very weak bonds with the mobile phase **(1)**.

2 This requires someone to print out at 100% and measure distance solvent front from start line (Y) and distance of spot from start line (middle)(X).

Then substitute the values for x and y in the following:

$Rf = X/y$ **(1)** = the answer **(1)**.

110. Investigating inks

1 (a) distance travelled by the spot / dye **(1)**; distance travelled by the solvent / solvent front **(1)**

(b) The measurements will be more precise / have a higher resolution **(1)** so the R_f value will be more accurate / closer to the true value **(1)**.

2 (a) to stop the solution boiling over (into the condenser) **(1)** so that the solvent collected is not contaminated with the solution / so that the vapour is not produced faster than it can be condensed **(1)**

(b) The apparatus will get very hot / solvent vapour (e.g. steam) will escape. **(1)**

3 (a) (highly) flammable **(1)**

(b) The student should work in a fume cupboard because the vapour causes dizziness **(1)**.

The propanone causes skin dryness, so the student should wear gloves / use forceps to handle the chromatogram **(1)**.

111. Drinking water

1 A **(1)**

2 (a) to sterilise the water / to kill microbes **(1)**

(b) The concentration of chlorine is high enough to kill microbes **(1)** but low enough so that it is not harmful to people **(1)**.

3 sedimentation **(1)** to remove larger insoluble particles **(1)**

filtration **(1)** to remove smaller insoluble particles **(1)**

4 Unlike tap water, distilled water does not contain dissolved salts **(1)**. These would interfere with the analysis / react with test substances / give a false-positive result **(1)**.

5 (a) (simple) distillation **(1)**

(b) A lot of energy is needed / a lot of fuel is needed to boil the water **(1)**.

6 $Al_2(SO_4)_3(aq) + 6H_2O(l) \rightarrow 2Al(OH)_3(s) + 3H_2SO_4(aq)$ **(1)**

112. Extended response – Separating mixtures

Answer could include the following points: **(6)**

Physical states:
- Substance A is solid; substances B and C are liquids.

Separating A from B and C:
- Substance A is insoluble in B and C,
- so it cannot be separated by paper chromatography
- but it can be separated from B and C by filtration.
- Substance A collects as a residue in the filter paper.
- It can be washed with B or C on the filter paper
- then dried in a warm oven
- below 115 °C to stop it melting.

Separating **B** and **C**:
- (After filtration) the filtrate is a mixture of substances **B** and **C**.
- They have different boiling points,
- so they can be separated by *fractional* distillation.
- Substance **B** has the lower boiling point.
- Substance **B** distils off first (and can be collected).
- Continue heating to leave substance **C** in the flask.
- Stop heating when the temperature starts to rise.

113. Acids and alkalis

1 D **(1)**

2 (a) The green colour means that the indicator is neutral **(1)**, so the pH is 7 **(1)**.

(b) red / orange **(1)**

3 (a) $2Mg + O_2 \rightarrow 2MgO$ **(1)**

(b) It is alkaline / contains an alkali / has a pH greater than 7 **(1)**.

4 **1 mark** for each correct row to **4 marks** maximum:

Formula of substance	Type of substance	
	Acid	Alkali
NaOH		✓
HCl	✓	
H_2SO_4	✓	
NH_3		✓

5 (a) litmus: blue, red **(1)**; phenolphthalein: pink, colourless **(1)**

(b) purple **(1)**

114. Bases and alkalis

1 B **(1)**

2 (a) sodium nitrate **(1)**

(b) sodium carbonate + nitric acid → sodium nitrate + water + carbon dioxide **(1)**

(c) bubbles **(1)**; powder disappears / dissolves / colourless solution forms **(1)**

3 (a) calcium chloride solution **(1)**

(b) hydrogen **(1)**

4 (a) (Bubble the gas through) limewater **(1)** which turns milky / cloudy white **(1)**.

(b) *Lighted* splint (ignites the gas) **(1)** with a (squeaky) pop **(1)**.

5 (a) A base is any substance that reacts with an acid **(1)** to form a salt and water only **(1)**.

(b) alkali **(1)**

(c) zinc sulfate **(1)**

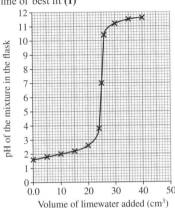

115. Neutralisation

1 Hydrogen ions, H^+, from the acid **(1)** react with hydroxide / OH^- ions from the alkali **(1)** to form water **(1)**.

2 (a) $CaO + 2HCl \rightarrow CaCl_2 + H_2O$ **(1)**

(b) $Ca(OH)_2 + 2HCl \rightarrow CaCl_2 + 2H_2O$

1 mark for formulae, **1 mark** for balancing

3 points plotted accurately (± ½ square) **(1)**; line of best fit **(1)**

Graph: pH of the mixture in the flask (y-axis, 0–12) against Volume of limewater added (cm^3) (x-axis, 0.0–50)

116. Salts from insoluble bases

1 (a) to react with *all* the acid **(1)** so that only a salt and water are left (with an excess of solid) **(1)**

(b) to make the reaction happen faster / to increase the rate of reaction **(1)**

(c) filtration / filtering **(1)**

(d) crystallisation / evaporation **(1)**

2 (a) measuring cylinder / pipette / burette **(1)**

(b) two improvements with reasons: **1 mark** for improvement, **1 mark** for its reason(s), e.g.
 - Stir, to mix the reactants.
 - Add copper oxide one spatula at a time, to reduce waste.
 - Warm the acid first, to make the reaction happen faster.

(c) Filter **(1)** to stop excess copper oxide contaminating the solution / crystals **(1)**.

117. Salts from soluble bases

1 C **(1)**

2 top label: burette **(1)**; bottom label: (conical) flask **(1)**

3 (a) hydrochloric acid **(1)**

(b) (volumetric) pipette **(1)**

(c) pink to colourless **(1)** *Both colours needed in the correct order for the mark.*

(d) (i) to get an idea of how much acid must be added **(1)**

(ii) run 1: 24.90; run 2: 24.40; run 3: 24.50 **(1)**

(iii) 24.90 cm^3 **(1)**

(iv) (24.40 + 24.50)/2 = 24.45 cm^3 **(1)**

(e) Repeat the titration without the indicator **(1)** using the mean titre volume of hydrochloric acid **(1)**.

118. Making insoluble salts

1 B **(1)**

2 D **(1)**

3 (a) calcium nitrate / calcium chloride **(1)** with sodium hydroxide / potassium hydroxide / ammonium hydroxide **(1)**

(b) Answer depends on the combination used in (a), e.g. sodium nitrate / potassium nitrate / ammonium nitrate (if calcium nitrate used); sodium chloride / potassium chloride / ammonium chloride (if calcium chloride used) **(1)**.

(c) The reaction is already very fast / the precipitate forms very quickly / the reaction has a low activation energy **(1)**.

4 (a) $Na_2CO_3(aq) + CaCl_2(aq) \rightarrow 2NaCl(aq) + CaCO_3(s)$

1 mark for balancing, **1 mark** for correct state symbols

(b) Add water to dissolve each solid separately / sodium carbonate and calcium chloride **(1)** then mix. Separate the precipitate of calcium carbonate using filtration **(1)**. Wash the precipitate using water (e.g. on the filter paper) **(1)**, then dry it by putting it in a warm oven / between pieces of filter paper **(1)**.

119. Extended response – Making salts

Answer could include the following points: **(6)**

The titration:
- Rinse a burette with dilute hydrochloric acid, then fill the burette with the acid.
- Measure 25 cm^3 of sodium hydroxide solution using a pipette into a conical flask on a white tile.
- Add a few drops of phenolphthalein / methyl orange indicator.
- Record the start reading on the burette.
- Add dilute hydrochloric acid from the burette to the sodium hydroxide solution.
- Swirl the flask.
- Rinse the inside of the flask during titration using distilled water.
- Add drop by drop near the end-point.
- Stop when colour changes / pink to colourless (phenolphthalein) / yellow to orange (methyl orange).
- Record the end reading on the burette.
- Repeat the experiment until consistent / concordant results are obtained.

Using the titre:
- Add 25 cm^3 of sodium hydroxide to the flask.
- Do not add indicator.
- Add the titre / mean titre volume of dilute hydrochloric acid from the burette.

Producing the crystals:
- Pour the mixture into an evaporating basin.
- Heat over a hot water bath
- until most of the water has evaporated.
- Allow to cool and pour away excess water.
- Dry crystals between filter paper / in a warm oven.

120. Electrolysis

1 D **(1)**

2 **1 mark** for each correct row to **4 marks** maximum:

	Positively charged	Negatively charged
Anode	✓	
Anion		✓
Cathode		✓
Cation	✓	

3 An electrolyte is an ionic **(1)** compound in the molten / liquid state or dissolved in water **(1)**.

4 bromine **(1)** *not bromide*

5 the negatively charged electrode, because sodium ions are positively charged / opposite charges attract **(1)** *Electrode and reason must be correct for the mark.*

6 MnO_4^- ions / manganate(VII) ions **(1)** move to the positively charged electrode / oppositely charged electrode **(1)**.

121. Electrolysing solutions

1 (a) B **(1)**

(b) D **(1)**

(c) copper **(1)**

2 (a) Na^+, Cl^- **(1)** H^+, OH^- **(1)**

(b) (i) chlorine **(1)**

(ii) hydrogen **(1)**

3 anode – oxygen **(1)**, cathode – hydrogen **(1)**

4 hydroxide / OH^- **(1)**

122. Investigating electrolysis

1 (a) inert **(1)**

(b) anode because it is positively charged **(1)** and oxygen is formed from negatively charged ions / hydroxide ions **(1)**

2 (a) time **(1)**

(b) gain in mass by copper cathode **(1)**

(c) non-inert because the anode loses mass / the cathode gains mass / the electrodes change in mass **(1)**

(d) change on the y-axis = 0.15 − 0.04 = 0.11 change on the x-axis = 0.8 − 0.2 = 0.6 **(1)**

gradient = 0.11/0.6 **(1)**

= 0.18 g/A **(1)**

123. Extended response – Electrolysis

Answer could include the following points: **(6)**

Solid copper chloride powder:
- Its ions are not free to move in the solid state so there are no visible changes and the solid powder does not conduct.

Copper chloride solution:
- Its ions are free to move when dissolved in water / in solution so the solution does conduct.
- Brown solid is copper.
- Yellow–green gas is chlorine.

Electrode reactions:
- positively charged ions / copper ions attracted to negative electrode / cathode

Answers

- negatively charged ions / chloride ions attracted to positive electrode / anode
- overall reaction: $CuCl_2(aq) \rightarrow Cu(s) + Cl_2(g)$

124. The reactivity series

1 (a) D (1)

(b) two from the following, for **1 mark** each: starting temperature (of water / acid); mass of metal; surface area of metal; amount / moles of metal

2 named metal that is less reactive than hydrogen, e.g. silver / copper / gold / platinum (1)

3 (a) (i) hydrogen (1)

(ii) *Lighted* splint ignites the gas (1) with a (squeaky) pop (1).

(b) magnesium oxide (1)

4 (a) $Al_2O_3(s) + 3H_2SO_4(aq) \rightarrow Al_2(SO_4)_3(aq) + 3H_2O(l)$ (1)

(b) To begin with, the acid reacts with the aluminium oxide layer (1) but, once this has gone, it reacts with the aluminium (1).

125. Metal displacement reactions

1 (a) Copper is more reactive than silver (1).

(b) $Cu(s) + 2AgNO_3(aq) \rightarrow 2Ag(s) + Cu(NO_3)_2(aq)$

1 mark for balancing, **1 mark** for state symbols

2 (a) copper (1)

(b) A metal cannot displace itself (1).

(c) (i) magnesium > metal X (1)

(ii) **1 mark** for experiment, **1 mark** for expected result, e.g. Put a piece of zinc into copper nitrate solution – if zinc is more reactive it gets coated / if zinc is less reactive there is no visible change **OR** Put a piece of copper in zinc nitrate – if copper is more reactive it gets coated / if copper is less reactive there is no visible change.

3 Aluminium is more reactive than iron (1) because aluminium can displace iron from its compounds / from iron oxide (1).

126. Explaining metal reactivity

1 A cation is a positively charged ion (1).

2 (a) Ca^{2+} (1)

(b) Two (1) electrons are lost from the outer shell (1).

(c) (i) potassium (1)

(ii) gold (1)

(d) copper / silver / gold (1)

3 Zinc is more reactive than copper (1) because it forms cations more easily / it loses electrons more easily (1).

127. Metal ores

1 C (1)

2 (a) A (1)

(b) Hydrogen is flammable / could explode (1).

3 a rock or mineral that contains enough metal / metal compound (1) to make its extraction worthwhile / economical (1)

4 (a) tin oxide + carbon → tin (1) + carbon monoxide / carbon dioxide (1)

(b) Carbon is oxidised (1) because it gains oxygen (1).

5 (a) Oil keeps air / oxygen away from the sodium (1).

(b) Copper is unreactive / low down on the reactivity series (1).

128. Iron and aluminium

1 (a) sodium / calcium / magnesium (1)

(b) lead / copper (1)

2 (a) iron oxide + carbon → iron (1) + carbon monoxide / carbon dioxide (1)

(b) Carbon is more reactive than iron / iron is less reactive than carbon (1).

(c) Iron oxide is reduced, because it loses oxygen (1).

3 (a) A lot of electricity is needed / electricity is more expensive than carbon (1).

(b) (i) aluminium (1)

(ii) oxygen (1)

(c) The carbon reacts with the oxygen formed / the electrodes burn away. (1)

4 Zinc is most likely to be extracted by heating zinc oxide with carbon (1) because zinc is less reactive than carbon / carbon is more reactive than zinc / electrolysis is more expensive than heating with carbon (1).

129. Recycling metals

1 **1 mark** for each correct row to **2 marks**:

Feature of recycling metals	Disadvantage (✓)
Used metal items must be collected.	✓
The use of finite resources is decreased.	
Different metals must be sorted.	✓
Metals can be melted down.	

Deduct 1 mark for each extra row completed.

2 two of the following, for **1 mark** each: dust produced; noisy; land used; wildlife loses habitat; extra traffic; landscape destroyed

3 (a) Most lead for recycling is found in batteries (1) so lead does not need to be sorted from scrap metal waste (1).

(b) two of the following, for **1 mark** each: conserves metal ores / limited resources; less energy needed; fewer quarries needed / saves land / landscape; less noise / dust produced

4 (a) aluminium (1)

(b) Steel and aluminium are much more abundant than tin in the Earth's crust (1); tin is much more valuable than steel or aluminium (1). *Accept the reverse arguments.*

130. Life-cycle assessments

1 2, 1, 4, 3 (1)

2 (a) (i) energy = 16.5 × 0.24 = 3.96 MJ (1)

(ii) 16.5 × 0.20 = 3.3 MJ (1)

(b) difference in mass between bottles in 1996 and 2016 = 0.24 – 0.20

= 0.04 kg (1)

difference in mass of CO_2 emitted = 0.04 × 1.2 = 0.048 kg (1)

3 (a) PVC: producing the material; wooden: transport and installation (1)

(b) the PVC frame (1) because it uses less energy / 20% of the energy / five times less energy (1)

131. The Haber process

1 (a) D (1)

(b) The reaction is reversible (1).

2 (a) temperature 450 °C (1); pressure: 200 atmospheres / 20 MPa (1)

(b) It is a catalyst (1); it makes the reaction happen faster (1).

3 (a) none / no visible change (1)

(b) (i) The rate of the forward and backward reactions is the same / equal (1) and they continue to happen (1).

(ii) They do not change / they remain constant (1).

132. Extended response – Reactivity of metals

Answer could include the following points: (6)

Basic method:

- I would start with powdered copper, iron, zinc, copper oxide, iron oxide and zinc oxide.
- Mix a spatula of a metal powder with a spatula of a metal oxide powder.
- Put the mixture in a steel lid.
- Heat strongly from below.
- Record observations.
- Repeat with a different combination of metal and metal oxide.

Expected results (in writing and / or as a table, as here):

	Copper oxide	Iron oxide	Zinc oxide
Copper	not done	no visible change	no visible change
Iron	reaction seen / brown coating	not done	no visible change
Zinc	reaction seen / brown coating	reaction seen / black coating	not done

Using the results:

- Count the number of reactions seen for each metal.
- Zinc has two reactions; iron has one reaction; copper has no reactions.
- order of reactivity (most reactive first): zinc, iron, copper

Controlling risks:

- Use tongs because substances / apparatus / steel lid is hot.
- Wear eye protection to avoid contact with (hot) powders.
- Stand back / use a safety screen / fume cupboard to avoid breathing in escaping substances / to avoid skin contact with hot powders.

133. The alkali metals

1 C **(1)**

2 table completed, e.g.

Alkali metal	Flame colour	Description
lithium	does not ignite	fizzes steadily
		disappears slowly
sodium	orange if ignited	fizzes rapidly **(1)**
		melts to form a ball / disappears quickly **(1)**
potassium	lilac **(1)**	fizzes very rapidly / gives off sparks **(1)**
		disappears very quickly / explodes at the end **(1)**

3 $2Na(s) + 2H_2O(l) \rightarrow 2NaOH(aq) + H_2(g)$

1 mark for balancing, **1 mark** for state symbols

4 Their atoms all have one electron in their outer shell **(1)**.

5 They are very reactive / react with water / react with air **(1)**; oil keeps water away / air away **(1)**.

6 (a) FrOH **(1)**

(b) Idea of a violent reaction, e.g. violent fizzing / explosion / flame / metal disappears almost immediately **(1)**.

7 Going down the Group, the size of the atoms increases **(1)**. The outer electron becomes further from the nucleus / more shielded **(1)** so the outer electron is lost more easily **(1)**.

134. The halogens

1 B **(1)**

2 Their atoms all have seven electrons in their outer shell **(1)**.

3 (a) chlorine: yellow–green **(1)** gas **(1)**; bromine: red–brown **(1)** liquid **(1)**

(b) very dark grey / black **(1)** solid **(1)**

4 (a) answer in the range 6000–7000 kg/m³ **(1)**

(b) Going down the Group, the density increases (and astatine is below iodine at 4933 kg/m³) **(1)**.

5 (a) covalent **(1)**

(b) intermolecular forces **(1)**

135. Reactions of halogens

1 (a) $H_2 + Cl_2 \rightarrow 2HCl$ **(1)**

(b) C **(1)**

(c) Going down Group 7, the elements become less reactive **(1)**. I can tell this

because the energy needed for them to start reacting increases (going down the Group / from fluorine to chlorine to bromine) **(1)**.

2 (a) sodium + chlorine → sodium chloride **(1)**

(b) FeCl₃ **(1)**

(c) (i) iron(II) ion: Fe²⁺ **(1)**; iodide ion: I⁻ **(1)**

(ii) Fe + I₂ → FeI₂

1 mark for correct formulae, **1 mark** for balancing

3 Fluorine atoms are smaller **(1)** than chlorine atoms, so its outer shell is closer to the nucleus **(1)** and it gains an outer electron more easily **(1)**.

136. Halogen displacement reactions

1 (a) D **(1)**

(b) bromine + potassium iodide → iodine + potassium bromide **(1)**

2 (a) The order of reactivity, starting with the most reactive, is chlorine, bromine, iodine **(1)** because chlorine displaces bromine from bromide and iodine from iodide **(1)** but bromine displaces only iodine from iodide. Iodine cannot displace chlorine or bromine **(1)**.

(b) A halogen cannot displace itself (so no reaction will be seen) **(1)**.

(c) Iodine is more reactive than astatine / astatine is less reactive than iodine **(1)** and a more reactive halogen will displace a less reactive halogen **(1)**.

3 (a) $2F_2(g) + 2H_2O(l) \rightarrow 4HF(aq) + O_2(g)$ **(1)**

(b) Fluorine reacts with water / does not form a solution of fluorine **(1)**.

(c) Chlorine is a pale (yellow–green) gas **(1)**.

137. The noble gases

1 A **(1)**

2 Balloons and airships rise because helium is less dense than air / has a low density **(1)**.

Helium is inert, so it will not catch fire **(1)**.

3 (a) helium **(1)**

(b) density increases down the Group **(1)**

(c) answer in the range −40 °C to −20 °C **(1)**; melting point increases down the Group **(1)**

4 The outer shells of their atoms are full / complete **(1)** so they have no tendency to gain / lose / share electrons / to form ions / to form covalent bonds **(1)**.

138. Extended response – Groups

Answer could include the following points: **(6)**

Reaction between sodium and chlorine:

- Sodium atoms transfer electrons from their outer shell to the outer shell of chlorine atoms.
- Each sodium atom loses one electron to form an Na⁺ ion.
- Each chlorine atom gains one electron to form a Cl⁻ ion.
- Oppositely charged ions / Na⁺ ions and Cl⁻ ions attract each other.
- Ionic bonds form.

- diagram showing the electronic configuration of Na, e.g. 2.8.1
- diagram showing the electronic configuration of Na⁺, e.g. 2.8 with + charge indicated
- diagram showing the electronic configuration of Cl, e.g. 2.8.7
- diagram showing the electronic configuration of Cl⁻, e.g. 2.8 with − charge indicated.

Violence of reaction between caesium and fluorine:

- Reactivity increases down Group 1 / Cs loses its electrons more easily (than Li, Na, K, Rb).
- Reactivity decreases down Group 7 / F gains electrons more easily (than any other Group 7 element).
- Caesium and fluorine are most reactive / very reactive.

139. Rates of reaction

1 **1 mark** for each correct row:

Change in reaction conditions	Frequency of collisions increased	Energy of collisions increased
increased concentration of a reacting solution	✓	
increased pressure of reacting a gas	✓	
increased temperature of reaction mixture	✓	✓

2 (a) It increases. **(1)**

(b) The powder has a larger *surface area : volume ratio* **(1)** so there are more frequent collisions between reactant particles **(1)**. *Not 'there are more collisions'.*

3 (a) a substance that speeds up a reaction without altering the products **(1)** and is unchanged chemically **(1)** and is also unchanged in mass **(1)** (at the end of the reaction)
The answers 'chemically' and 'in mass' can be in either order.

(b) A catalyst provides an alternative pathway **(1)** with a lower activation energy **(1)**.

(c) (i) enzyme **(1)**

(ii) making alcoholic drinks / wine / beer **(1)**

140. Investigating rates

1 (a) Collect the gas in a gas syringe / upturned burette of water / upturned measuring cylinder of water **(1)**.

(b) The sulfur dioxide will dissolve in the water **(1)** so most will not escape / the volume measurement will be too low **(1)**.

(c) Sodium chloride solution and water are both clear and colourless **(1)** so you could not tell that they are being produced **(1)**.

2 (a) Total volume of liquid is kept the same **(1)**; concentration of dilute hydrochloric acid is kept the same **(1)**.

(b) 8 **(1)**, 24 **(1)**, 40 **(1)**

(c) As the concentration increases the rate increases / rate is proportional to concentration (1); when the concentration increases three times the rate increases three times / when the concentration increases five times the rate increases five times (1). *Or* Rate is *directly* proportional to the concentration (2).

141. Exam skills – Rates of reaction

1 (a) all points plotted correctly ± ½ square (2)
   1 mark if one error
   single line of best fit passing through all the points (1)
   Do not use a ruler to join the points (apart from the last two) because a curve is required here.
  (b) time taken: 100 s (1)
   Explanation: the mass does not change any more / the line becomes horizontal. (1) *not 'straight'*
  (c) line drawn to the left of the original line (1) becoming horizontal at 0.96 g (1)

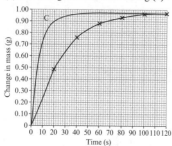

142. Heat energy changes

1 In an exothermic change or reaction, heat energy is given out (1) but, in an endothermic change or reaction, heat energy is taken in (1).

2 C (1)

3 (a) endothermic (1)
  (b) acid–alkali neutralisation / aqueous displacement (1)

4 (a) Measure the temperature of the acid before and after adding magnesium (1) using a thermometer (1) and the temperature should increase (1).
  (b) $Mg(s) + 2HCl(aq) \rightarrow MgCl_2(aq) + H_2(g)$
   1 mark for balancing, **1 mark** for state symbols
  (c) More heat energy (1) is released when bonds form in the products (1) than is needed to break bonds in the reactants (1).

143. Reaction profiles

1 activation energy (1)

2 There is more stored energy in the reactants than in the products (1) so, during the reaction, energy is given out (1).

3 diagram completed with upwards arch between reactants and product lines (1); activation energy correctly identified (1)

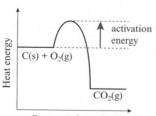

4 activation energy correctly identified (1); overall energy change correctly identified (1)

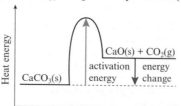

144. Crude oil

1 (a) D (1)
  (b) D (1)
  (c) It is no longer being made / it is made extremely slowly (1).

2 (a) C_6H_{12} (1)
  (b) They are compounds of hydrogen and carbon (1) only (1).

3 (a) petrol / diesel oil / kerosene / fuel oil (1)
  (b) a starting material (1) for an industrial chemical process (1)

145. Fractional distillation

1 (a) (i) gases (1)
     (ii) bitumen (1)
     (iii) gases (1)
     (iv) bitumen (1)
  (b) (i) bitumen (1)
     (ii) kerosene (1)
  (c) petrol (1), diesel oil (1)

2 alkanes (1)

3 Oil is heated so that it evaporates / boils / vaporises (1). The vapours are passed into a column, which is hot at the bottom and cold at the top (1).

   Hydrocarbons rise, and condense (1) at different heights, depending on boiling point / size of molecules / strength of intermolecular forces (1).

146. Alkanes

1 C (1)

2 A (1)

3 (a) C_2H_6 (1)
  (b) Correct structure for **1 mark**, e.g.

  (c) Going from one alkane to the next, the molecular formula changes by CH_2 / one carbon atom *and* two hydrogen atoms. (1)

4 (a) C_nH_{2n+2} (1)
  (b) C_6H_{14} (1)

147. Incomplete combustion

1 (a) complete combustion (1)
  (b) **1 mark** for each correct column, e.g.

	Incomplete combustion	Complete combustion
Water	✓	✓
Carbon	✓	
Carbon monoxide	✓	
Carbon dioxide	✓	✓

2 flame **A** because incomplete combustion occurs (1) producing soot / carbon particles (1)

3 (a) They cause lung disease / bronchitis / make existing lung disease worse, e.g. asthma (1).
  (b) $C_6H_6 + 4O_2 \rightarrow 2C + 3CO + CO_2 + 3H_2O$ (1)
  (c) When breathed in, carbon monoxide combines with haemoglobin / red blood cells (1) so less oxygen can be carried / there is a lack of oxygen to cells (1).

148. Acid rain

1 (a) clockwise: rain cloud (1); acid rain (1); distant city (1) power station (1); acidic gases (1);
  (b) Sulfur (impurities) in the fuel (1) react with oxygen (1).

2 Oxygen and nitrogen from the air (1) react together at the high temperatures inside the engine (1).

3 (a) Marble / calcium carbonate reacts with acids / acidic rainwater (1) but granite does not (1).
  (b) Damage to trees / plants / soil (1) makes lakes acidic / harms aquatic life (1).

149. Choosing fuels

1 (a) crude oil (1)
  (b) Petrol is used as a fuel for cars. (1) Kerosene is used as a fuel for aircraft (1) and diesel oil is used as a fuel for some cars / some trains (1).

2 It is being used up faster than it can form (1).

3 (a) $2H_2 + O_2 \rightarrow 2H_2O$ (1)
  (b) carbon dioxide (1)

4 (a) (i) Petrol releases more energy per dm^3 than hydrogen does (1).
     (ii) Hydrogen releases more energy per kg than petrol does (1).
  (b) More liquid hydrogen can be stored in the same volume / same number of particles in the liquid can be stored in a smaller volume (1).

150. Cracking

1 (a) alkene / unsaturated (1)
  (b) $C_{10}H_{22} \rightarrow C_8H_{18} + C_2H_4$
   1 mark for correct reactant, **1 mark** for correct products

2 D (1)

3 (a) It is a catalyst **(1)**.

(b) a reaction in which larger alkanes are broken down into smaller (more useful) alkanes / smaller (more useful) saturated hydrocarbons **(1)** and smaller alkenes / unsaturated hydrocarbons **(1)**

4 Smaller hydrocarbons are in higher demand / more useful **(1)**. Cracking helps to match supply with demand **(1)**.

151. Extended response – Fuels

Answer could include the following points: **(6)**

Why incomplete combustion happens:

- Incomplete combustion happens when there is insufficient oxygen / air.
- This can happen if there is not enough ventilation, such as inside a tent.
- Not enough oxygen for complete combustion.

Products and their problems:

- Carbon monoxide gas produced.
- Carbon monoxide is toxic.
- Carbon monoxide is colourless / odourless so it may not be noticed.
- Combines with haemoglobin / red blood cells so less oxygen can be carried / there is a lack of oxygen to cells.
- Can cause unconsciousness / death.
- Carbon particles / soot produced.
- Cause lung disease / bronchitis / make existing lung disease worse.
- Cause blackening, e.g. of the bottom of the kettle / the inside of the tent.
- Balanced equation, e.g. $C_3H_8 + 3O_2 \rightarrow 4H_2O + 2CO + C$. *Many incomplete combustion reactions and their equations are possible.*

Other problems:

- Less energy is released by incomplete combustion.

152. The early atmosphere

1 B **(1)**

2 (a) 10% **(1)**

(b) Water vapour (in the atmosphere) **(1)** condensed (and fell as rain) **(1)**.

(c) Carbon dioxide dissolved **(1)** in the oceans / water **(1)**.

3 (a) A *glowing* splint **(1)** relights **(1)**.

(b) The growth of primitive plants used carbon dioxide **(1)** and released oxygen **(1)** by the process of photosynthesis **(1)**.

153. Greenhouse effect

1 B **(1)**

2 correct order (top to bottom of the table): 2, 4, 1, 3

1 mark for two correct numbers; **3 marks** for all four numbers correct

3 (a) methane **(1)**

(b) Fossil fuels contain hydrocarbons / carbon **(1)**, which react with oxygen in the air to produce carbon dioxide **(1)**.

4 (a) It increases **(1)**.

(b) a worldwide increase **(1)** in temperatures **(1)**

(c) two from the following, for **1 mark** each: climate change / change in global weather patterns / ice caps melting / sea levels rise / loss of habitats

154. Extended response – Atmospheric science

Answer could include the following points: **(6)**

Greenhouse effect:

- Carbon dioxide and some other gases in the atmosphere absorb heat energy radiated from the Earth then release energy which keeps the Earth warm.

Processes releasing carbon dioxide:

- burning fossil fuels
- respiration
- volcanic activity.

Processes absorbing carbon dioxide:

- dissolving in sea water
- photosynthesis.

Discussing the data:

- As the concentration of carbon dioxide rises the mean global temperature rises.
- Human activity, e.g. burning fossil fuels, could cause an increase in temperature but there are some years when the temperature decreases, e.g. in the late 1940s.
- Carbon dioxide is also produced by other processes, e.g. by volcanic activity so it might not all be due to human activity.
- There might be a common factor not shown in the graphs that is responsible for both changes.

Physics

155. Key concepts

1 ampere, A; joule, J; pascal, Pa; watt, W; newton, N; ohm, Ω (all six for **2 marks**, any three or four for **1 mark**)

2 A base unit is independent of any other unit **(1)**; a derived unit is made up from two or more base units **(1)**.

3 (a) 0.75 kg **(1)** (b) 750 W **(1)** (c) 1500 s **(1)**

4 (a) 8 nm = 0.000 000 008 m **(1)**

(b) 8×10^{-9} m **(1)**

5 $s = d \div t = 75$ m ÷ 10.5 s = 7.142 857 m/s **(1)** so to 5 significant figures = 7.1430 m/s **(1)**

156. Scalars and vectors

1 (a) scalars: speed, temperature, mass, distance **(1)**; vectors: acceleration, displacement, force, velocity **(1)**

(b) (i) any valid choice and explanation, e.g. mass **(1)** is a scalar because it has a size / magnitude but no direction **(1)**

(ii) any valid choice and explanation, e.g. weight **(1)** is a vector because it has a magnitude and direction **(1)**

2 (a) Velocity is used because both a size **(1)** and a direction **(1)** are given

(b) The second swimmer is swimming in the opposite direction to the first swimmer **(1)**.

3 (a) D **(1)**

(b) Weight has a size / magnitude and a direction but all the other quantities just have a magnitude **(1)**.

4 (a) C: –3 m/s; A: 5 m/s; B: 4.5 m/s (accept reverse for A and B) (all three correct for **2 marks**; two correct for **1 mark**)

(b) Runner C is running in the opposite direction so the velocity is given a negative sign **(1)**.

157. Speed, distance and time

1 (a) (i) B **(1)**

(ii) C **(1)**

(b) In part A he travels 60 m in 40 s. Speed = distance ÷ time **(1)** = 60 m ÷ 40 s **(1)** = 1.5 m/s **(1)**. (You can use any part of the graph to read off the distance and the time as the line is straight; you should always get the same speed.)

(c) A faster speed is shown on the graph by a steeper gradient **(1)** of the distance / time line; a slower speed is shown on the graph by a less steep gradient **(1)** (comparison must be made for second mark).

2 (a) speed = 84 m ÷ 24 s **(1)** = 3.5 **(1)** m/s **(1)**

(b) 3.5 m/s upwards or up **(1)**

3 time = distance ÷ speed = 400 m ÷ 5 s **(1)** = 80 s **(1)**

158. Equations of motion

1 C **(1)**

2 *a*: acceleration; *t*: time; *x*: distance; *u*: initial velocity (all four correct for **2 marks**; two correct for **1 mark**)

3 (a) $a = (v - u) \div t$ **(1)** = (25 m/s − 15 m/s) ÷ 8 s **(1)** = 1.25 m/s² **(1)**

(b) $v^2 = u^2 + 2 \times a \times x$ = (25 m/s)² + 2(1.25 m/s² × 300 m) **(1)** = 1375 m²/s² **(1)**
$v = \sqrt{1375}$ m/s = 37 m/s **(1)**

159. Velocity/time graphs

1 (a) A **(1)**

(b) a right-angled triangle with a horizontal and vertical side that covers as much of the line as possible **(1)**.

(c) change in velocity = 30 m/s, time taken for change = 5 s; acceleration = (change in velocity) ÷ (time taken) = 30 m/s ÷ 5 s **(1)** = 6 m/s² **(1)** (the triangle drawn may be different but the answer should be the same)

(d) area under line = ½ × 5 s × 30 m/s **(1)** = 75 m **(1)**

2 (a) $a = (v - u) \div t$ = (4 − 0) ÷ 5 **(1)** = 0.8 **(1)** m/s² **(1)**

(b) Plot a velocity / time graph **(1)**; the total area under the line can be calculated **(1)** and this gives the total distance travelled **(1)**.

160. Determining speed

1 commuter train: 55 m/s; speed of sound in air: 330 m/s; walking: 1.5 m/s **(1)** (all three needed for **1 mark**)

2 (a) The light beam is cut / broken by the card as it enters the light gate and this starts the timer **(1)**. When the card has passed through the light beam is restored **(1)** and this stops the timer **(1)**.

(b) Speed is calculated from the length of the card and the time taken for the card to pass through the light gate **(1)**.

3 Very short distances may be measured using this method and this gives a good measure

of instantaneous speed **(1)**. Short times are difficult to measure accurately with a stopwatch **(1)**. Errors associated with human error / reaction time / parallax are reduced **(1)**.

161. Newton's first law

1 A resultant force determines the movement of an object. **(1)** It is the single net force remaining after adding together all the forces acting on an object **(1)**.

2 (a) The action is the downwards force of the skater on the ice **(1)** and the reaction is the (upwards) force of the ice on the skater **(1)**.
   (b) resultant force = 30 N − 10 N − 1 N **(1)** = 19 N **(1)**
   (c) The resultant force is zero / 0 N **(1)** so the velocity is constant / stays the same **(1)**.
   (d) Only the resistance forces are acting now, so the resultant force is backwards / against the motion of the skater **(1)**, so the skater slows down / has negative acceleration **(1)**.

3 (a) Assume downwards is positive; so resultant downward force = 1700 N − 1900 N = − **(1)** 200 N **(1)** (State which direction you are using as the positive direction.)
   (b) The velocity of the probe towards the Moon will decrease **(1)** because the resultant force acts upwards / is negative **(1)**.

162. Newton's second law

1 (a) The trolley will accelerate **(1)** in the direction of the pull / force **(1)**.
   (b) The acceleration is smaller / lower **(1)** because the mass is larger **(1)**.

2 (a) $F = m \times a = 3000$ N $\times 39$ m/s^2 **(1)** = −117 000 **(1)** N **(1)**
   (b) in the opposite direction to the spacecraft's motion / upwards **(1)**

3 (a) $a = F \div m = 10\,500$ N $\div 640$ kg **(1)** = 16.4 **(1)** m/s^2 **(1)**
   (b) The mass of the car decreases **(1)** so the acceleration will increase **(1)**.

163. Weight and mass

1 A **(1)**

2 The mass of the LRV on the Moon is 210 kg **(1)** because the mass of an object does not change if nothing is added or removed **(1)**.

3 (a) $W = m \times g$ **(1)** so (1 + 2 + 1.5) kg × 10 N/kg = 4.5 kg × 10 N/kg **(1)** = 45 N **(1)**
   (b) $m = W \div g$ so 30 ÷ 10 = 3 kg **(1)**, 3 kg = mass of bag (1 kg from (a)) + mass of sports kit so mass of sports kit will be 3 − 1 = 2 kg **(1)**

164. Force and acceleration

1 Electronic equipment is much more accurate **(1)** than trying to obtain accurate values for distance and time to calculate velocity, then calculate acceleration **(1)**, when using a ruler and a stopwatch. (Reference should be made to distance, time and velocity.)

2 For the same force, larger mass will result in lower acceleration, smaller mass will result in higher acceleration **(1)**.

3 Acceleration is the change in speed ÷ time taken so two velocity values are needed **(1)**; the time difference between these readings **(1)**

is used to obtain a value for the acceleration of the trolley.

4 For a constant slope, as the mass increases, the acceleration will decrease **(1)** due to greater inertial mass **(1)**.

5 any one of the following hazards: An accelerating mass of greater than a few hundred grams can be dangerous and may hurt somebody if it hits them at speed **(1)**; electrical apparatus could cause a shock if not handled correctly / bare wires are used / the apparatus is faulty **(1)**; trailing electrical leads could cause a fall so these should be carefully arranged out of the way **(1)**. (Accept any sensible suggestion.)

   any two of the following precautions: Do not use masses greater than a few hundred grams **(1)**, wear eye protection **(1)**, use electrically tested electronic equipment **(1)**, avoid trailing electrical leads **(1)**. (Accept any sensible suggestion.)

165. Newton's third law

1 D **(1)**

2 As the rocket motors fire up and push out the hot gases downwards **(1)**, the hot gases push upwards on the rocket **(1)**. As the upward push / thrust on the rocket becomes greater than the weight of the rocket, the rocket lifts off **(1)**.

3 (a) As the athlete pushes on the brick wall, the wall pushes back **(1)** with an equal and opposite force **(1)**.
   (b) The curtain is not rigid and so would not push back with an equal and opposite force **(1)**. If the athlete used the same force as she used on the brick wall the forces would be unbalanced and the athlete (pushing with a larger force) would fall forward **(1)**.

166. Human reaction time

1 (a) B **(1)**
   (b) any one of the following: tiredness / alcohol / drugs / distractions **(1)**

2 Human reaction time is the time between a stimulus occurring and a response **(1)**. It is related to how quickly the human brain can process information and react to it **(1)**.

3 The person taking the test sits with their index finger and thumb opened to a gap of about 8 cm. A metre ruler is held, by a partner, so that it is vertical and exactly level with the person's finger and thumb, with the lowest numbers on the ruler at the bottom. The ruler is dropped and then grasped by the person taking the test. The position of the grab on the ruler is noted and the experiment is repeated. (any three from the four answers given for **3 marks**)

4 any suitable example **(1)** with justification **(1)** (e.g. international tennis player who is able to react to the opponent hitting the ball; racing driver who can operate a vehicle at very high speeds, airline pilot who can react to avoid unexpected aircraft)

167. Stopping distance

1 thinking distance + braking distance = overall stopping distance **(1)**

2 Thinking distance will increase if: the car's speed increases, the driver is distracted, the

driver is tired, the driver has taken alcohol or drugs **(1)** (all four needed for **1 mark**). Braking distance will increase if: the car's speed increases, the road is icy or wet, the brakes or tyres are worn, the mass of the car is bigger **(1)** (all four needed for **1 mark**).

3 Worn tyres will have less tread so water remains between the tyre and the road resulting in less contact **(1)**, therefore friction is reduced and a car with worn tyres takes longer to stop **(1)**. (This affects the braking distance component. The thinking distance component remains unchanged.)

4 When Dom drives alone, the total mass **(1)** of the vehicle will be lower **(1)** than when his family and their luggage are in the car. For the same speed, **(1)** Dom will need to allow a longer stopping distance **(1)** with his family in the car than when he was travelling alone **(1)**.

168. Extended response – Motion and forces

Answer could include the following points: **(6)**

- As the skydiver jumps from the plane, weight / force acts in a downward direction.
- The upward / resistive force is caused by air resistance acting on the skydiver.
- The downward and upward forces act in opposite directions to give a resultant force.
- The resultant force is downwards / causing the skydiver to accelerate.
- The 'spread-out' position of the skydiver increases the surface area exposed to the air resistance, increasing the upward force and reducing the rate of acceleration.
- Maximum velocity occurs when the downward and upward forces are in balance.
- The parachute increases air resistance, increasing the upward force.
- Increased upward force causes the skydiver to decelerate.
- When the skydiver and parachute have slowed to a velocity where the upwards and downwards forces are again balanced, the skydiver falls to the ground at a slower steady velocity.

169. Energy stores and transfers

1 D **(1)**

2 An energy transfer diagram shows both the input energy into a system **(1)** and the output energy **(1)** by using a large arrow to represent the movement of energy **(1)**.

3 chemical store **(1)**; kinetic store and thermal store **(1)**; thermal store **(1)**

4 Gravitational store bar is lower **(1)**; kinetic store bar is higher **(1)**; rise in kinetic store bar should equal the amount down in the gravitational store bar **(1)**.

170. Efficient heat transfer

1 concrete **(1)**: this has the lowest relative thermal conductivity, which means that it will have the slowest rate of transfer of thermal energy **(1)**

2 Thicker walls provide more material for the thermal energy to travel through from the inside to the outside **(1)** so the rate of thermal energy loss is less, keeping the houses warmer **(1)**.

3 efficiency = useful energy transferred by the crane ÷ total energy supplied to the crane; useful energy transferred = energy transferred to the box = 1 000 000 J; total energy used by the crane = the energy stored in the fuel = 4 000 000 J; efficiency = 1 000 000 J ÷ 4 000 000 J × 100 **(1)** = 25% **(1)** (or calculation can omit × 100 and leave efficiency as 0.25)

4 (a) efficiency = 20%; if motor transfers 40 J/s to the kinetic energy store (useful energy) then 20% = 1/5 total energy transferred to the motor **(1)** so to calculate 100%, 5 × 40 **(1)** = 200J **(1)**
   (b) 200 W **(1)**

171. Energy resources

1 coal, oils and gas **(1)** (all three needed for mark)

2 (a) hydroelectric **(1)**; (b) tidal **(1)**; wind **(1)**

3 (a) hydroelectric and geothermal **(1)**
   (b) Demand is greatest at certain times of the day **(1)**. Demand may be high when some renewable sources may not be available **(1)**.

4 Geothermal and oil resources both heat water to produce steam to drive turbines **(1)**. Geothermal and oil power stations are not dependent on weather conditions to drive turbines **(1)**. Geothermal and oil power stations both use resources in the Earth's crust **(1)**.

5 (a) A hydroelectric power station is a reliable producer of electricity because it uses the gravitational potential energy of water which can be stored until it is needed **(1)**. As long as there is no prolonged drought / lack of rain the supply should be constant **(1)**.
   (b) any of the following: Hydroelectric power stations have to be built in mountainous areas / high up (compared with supply areas so that the gravitational potential energy can be captured) **(1)**. The UK has very few mountainous areas like this **(1)**. It is limited to areas such as north Wales and the Scottish Highlands **(1)**.

172. Patterns of energy use

1 (a) After 1900 the world's energy demand rose as the population grew **(1)**. There was development in industry which increased the demand for energy **(1)** and the rise of power stations using fossil fuels added to demand **(1)**.
   (b) (i) non-renewable energy resources / fossil fuels **(1)**
      (ii) any two from; Population has increased so energy consumption is higher **(1)**, industrial / technological developments require more energy **(1)**, transport networks have grown **(1)** (any other valid reason)
      (iii) Nuclear energy research did not begin until after 1900 **(1)**.
      (iv) hydroelectric **(1)**

2 any four of the following points: As the population continues to rise the demand for energy will also continue to rise **(1)**. Current trends show the use of fossil fuels being a major contributor to the world's energy resources **(1)**. These are running out and no other energy resource has so far taken their place **(1)**. This could lead to a large gap between demand and supply **(1)**. To match the rise in demand for energy, further research and development of non-renewable resources will need to be made **(1)** to provide reliable **(1)** and cost-effective **(1)** energy supplies. While cheaper fossil fuels still remain available there is less incentive for governments to do this **(1)**.

173. Potential and kinetic energy

1 (a) When an object is raised up above the Earth's surface **(1)** it gains gravitational potential energy.
   (b) mass **(1)**, height **(1)**

2 D **(1)**

3 ΔGPE = $m \times g \times \Delta h$ = 2 kg × 10 N/kg × 25 m **(1)** = 500 **(1)** J **(1)**

4 left: B or C; middle: A; right: B or C (all three correct – **2 marks**, two correct – **1 mark**)

5 kinetic energy = $\frac{1}{2} \times m \times v^2$, so KE = $\frac{1}{2} \times 70 \times 6^2$ **(1)** = 1260 **(1)** J

174. Extended response – Conservation of energy

Answer could include the following points: **(6)**

● Refer to the change in gravitational potential energy (GPE) as the swing seat is pulled back / raised higher.
● Before release, the GPE is at maximum / kinetic energy (KE) of the swing is at a minimum.
● When the swing is released, the GPE store falls and the KE store increases.
● KE is at a maximum at the mid-point; GPE is at a minimum.
● The system is not 100% efficient; some energy is dissipated to the environment.
● Friction due to air resistance and / or at the pivot results in the transfer of thermal energy to the surroundings / environment.
● Damping, due to friction, will result in the KE being transferred to the thermal energy store of the swing / environment.
● Eventually all the GPE will have been transferred to other energy stores and will no longer be useful.

175. Waves

1 Sound waves are this type of wave; L. All electromagnetic waves are this type of wave: T. Particles oscillate in the same direction as the wave: L. They have amplitude, wavelength and frequency: B. Seismic P waves are this type of wave: L. They transfer energy: B. (All six correct – **3 marks**; five correct – **2 marks**; three correct – **1 mark**.)

2 (a) B **(1)**
   (b) 6 cm/0.06 m **(1)**
   (c) any correct wave with higher amplitude **(1)** and shorter wavelength **(1)**

3 When a sound wave moves through the air the particles oscillate **(1)** in the same direction as the direction in which the wave travels **(1)**.

176. Wave equations

1 (a) speed = m/s; distance = m; time = s (all 3 correct – **2 marks**; 2 correct – **1 mark**.
   (b) speed = distance / time **(1)** (accept alternative arrangements of distance = speed × time or time = distance / speed)

2 distance travelled by the waves (in metres) = 30 000 m **(1)**; time taken = 20 s; average speed of sound = 30 000 m ÷ 20 s **(1)** = 1500 m/s **(1)**

3 (a) $v = f \times \lambda$ or wave speed = frequency × wavelength **(1)**
   (b) wave speed = 0.017 m × 20 000 Hz **(1)** = 340 m/s **(1)**

4 $\lambda = v \div f$ = 0.05 m/s ÷ 2 **(1)** = 0.025 **(1)** m **(1)**

177. Measuring wave velocity

1 frequency of the waves (f) = 3 Hz; wavelength of the waves (λ) = 0.05 m; speed of waves = 3 Hz × 0.05 m **(1)** = 0.15 **(1)** m/s **(1)**

2 D **(1)**

3 time taken to reach the fish: t = x / v so t = 150 m / 1500 m/s **(1)** so t = 0.1 s **(1)** so time taken for the reflected sound to come back to the dolphin = 2 × 0.1 s = 0.2 s **(1)**

4 two from: Use an electronic data collector **(1)**; repeat the experiment at 50 m **(1)**; repeat the experiment over a range of distances **(1)**.

178. Waves and boundaries

1 D **(1)**

2 (a)

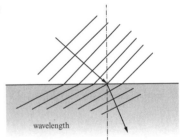

wavelength

Normal should be drawn at 90° to the boundary between the two materials **(1)**. (b) The ray should be drawn 'bending' towards the normal in the second material **(1)**. (c) Wave fronts should be drawn perpendicular to the ray in the second material (judge by eye) – at least three wave fronts drawn **(1)**. Wave fronts should be closer together than in material 1 **(1)**.

3 (a) The wave bends away from **(1)** the normal.
   (b) The wave bends towards **(1)** the normal.

4 (a) The light ray should not change direction at the boundary between the air and glass **(1)**. The ray should continue through the glass and not change direction at the boundary between the glass and air **(1)**.

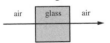

air   glass   air

179. Waves in fluids

1 (a) Count the number of waves that pass a point each second and do this for one minute **(1)**; divide the total by 60 to get a more accurate value for the frequency of the water waves **(1)**.
   (b) Use a stroboscope to 'freeze' the waves **(1)** and find their wavelength by using a ruler in the tank / on a projection **(1)**.
   (c) wave speed = frequency × wavelength or $v = f \times \lambda$ **(1)**
   (d) the depth of the water **(1)**

2 A ripple tank can be used to determine the three values for wavelength, frequency and wave speed of water waves **(1)**, as long as small wavelengths **(1)** and small frequencies **(1)** are used.

3 water: hazard – spills may cause slippages; safety measure – report and wipe up immediately **(1)**; electricity: hazard – may cause shock or trailing cables; safety measure – do not touch plugs / wires / switches with wet hands or keep cables tidy **(1)**; strobe lamp: hazard – flashing lights may cause dizziness or fits; safety measure – check that those present are not affected by flashing lights **(1)**

180. Extended response – Waves

Answer could include the following points: **(6)**

- The phenomenon is refraction, a property of waves.
- Light is a wave and so is refracted through transparent or translucent materials.
- When light passes from one material to another of different density it is refracted at the boundary and changes direction.
- When light passes from a more dense material to a less dense material it is refracted away from the normal (as in this case).
- The actual position of the key is at position A.
- The key appears to be at position B because the man's brain extrapolates the refracted wave (as shown by the dashed line).

181. Electromagnetic spectrum

1 A **(1)**

2 B **(1)**

3 (a) A: X-rays **(1)**; B: visible light **(1)**; C: microwaves **(1)**

(b) The different waves carry different amounts of energy **(1)**.

4 $v = f \times \lambda$, so $f = v \div \lambda$ **(1)** = 3×10^8 m/s ÷ 240 m **(1)** = 1.25×10^6 Hz **(1)**

182. Investigating refraction

1 (a) Place a refraction block on white paper and connect a ray box to an electricity supply **(1)**; switch on the ray box and set it at an angle to the surface of the block **(1)**; use a sharp pencil to draw around the refraction block and make dots down the centre of the rays either side of the block **(1)**; use a sharp pencil and ruler to join the 'external' rays and then draw a line across the outline of the block to join the lines **(1)**; use a protractor to draw a normal where the light ray met the block and measure the angle of incidence and angle of refraction **(1)**.

(b) When a light ray travels from air into a glass block, its direction changes **(1)** and the angle of refraction will be less than the angle of incidence **(1)**.

(c) The ray of light would not change direction **(1)**.

2 any three from: use of electricity: if mains electricity is used there is a risk of shock – use tested apparatus / do not try to plug in / unplug in the dark **(1)**; experiments

are generally done in low-level light so there is a risk of tripping – clear floor area and working space (no trailing wires) and avoid moving around too much **(1)**; if glass blocks are used there is a risk of cuts – handle with care, use Perspex/non-breakable blocks when possible **(1)**; ray boxes get hot so risk of burns – do not touch ray boxes during operation **(1)**.

3 Transparent materials allow light to pass through and so are good for investigating refraction **(1)**. Translucent materials allow some light through but tend to scatter light so measuring refraction is difficult **(1)**. Opaque materials reflect light and so are unsuitable for investigating refraction **(1)**.

183. Dangers and uses

1 (a) frequency **(1)**

(b) A – 3; B – 1; C – 4; D – 2 (all correct – **2 marks**; 3 correct – **1 mark**; 2 or less – **0 marks**)

2 (a) A and C **(1)**

(b) B and D **(1)**

(c) A and D **(1)**

3 any valid uses and expected location, e.g. observing the internal structure of non-living objects / underground pipes **(1)**; security scanners / airports **(1)**; medical diagnostics / hospitals **(1)**

4 (a) Microwaves cause damage to cells **(1)** by causing internal heating **(1)**.

(b) Infrared waves transfer thermal **(1)** energy and can cause burns **(1)**.

(c) Ultraviolet waves can damage eyes **(1)** and can cause skin cancer **(1)**.

184. Changes and radiation

1 D **(1)**

2 (a) When an electron absorbs electromagnetic radiation it moves up one or more energy levels in the atom or may leave the atom entirely **(1)**.

(b) When an electron emits electromagnetic radiation it moves down one or more energy levels in the atom **(1)**.

3 C **(1)**

185. Extended response – Light and the electromagnetic spectrum

Answer could include the following points: **(6)**

- X-rays and gamma-rays are both transverse waves.
- X-rays and gamma-rays have high frequency and therefore carry high amounts of energy.
- X-rays and gamma-rays cause ionisation in atoms and exposure can be dangerous / cause cells to become cancerous.
- People who work regularly with X-rays and gamma-rays should limit their exposure by using shields or by leaving the room during use.
- Low-energy X-rays are transmitted by normal body tissue but are absorbed by bones and other dense materials such as metals.
- Higher-energy gamma-rays mostly pass through body tissue but can be absorbed by cells.
- X-rays and gamma-rays can be used to investigate / treat medical problems.

- X-rays can be used to 'see' inside containers, e.g. at border controls or at airport security / to check cracks in metals.
- Gamma-rays can be used to kill bacteria in food / on surgical instruments.

186. Structure of the atom

1 B **(1)**

2 (a) protons labelled in the nucleus (+ charge) **(1)**

(b) neutrons labelled in nucleus (0 charge) **(1)**

(c) electrons labelled as orbiting (− charge) **(1)**

3 (a) The number of positively charged protons in the nucleus **(1)** is equal to the number of negatively charged electrons orbiting the nucleus **(1)**.

(b) The atom will become a positively charged ion / charge of +1 **(1)**.

4 (a) A molecule is two or more atoms bonded together. **(1)** (In the kinetic theory of gases, molecule also describes monatomic gases).

(b) (i) any pure liquid, e.g. water / H_2O **(1)**

(ii) any gaseous molecule, e.g. oxygen / O_2 **(1)**

(iii) any gaseous compound, e.g. carbon dioxide / CO_2 **(1)**

187. Atoms and isotopes

1 (a) the name given to particles in the nucleus **(1)**

(b) the number of protons in the nucleus **(1)**

(c) the total number of protons and neutrons in the nucleus **(1)**

2 (a) All atoms of the same element have the same number of protons in their nuclei **(1)**.

(b) An isotope is an atom of the same element that has the same number of protons **(1)** but a different number of neutrons **(1)**.

3 B **(1)**

4 Isotopes will be neutral because neutrons have no charge **(1)** so changing the number of neutrons does not affect the balance of positive and negative charges of the atom **(1)**.

5 $^{12}_{6}$C, 6 protons, 6 neutrons **(1)**; $^{35}_{17}$Cl, 17 protons, 18 neutrons **(1)**; $^{23}_{11}$Na, 11 protons, 12 neutrons **(1)** (all three in each line needed for each mark)

188. Atoms, electrons and ions

1 (a) 5 **(1)**

(b) The diagram shows 5 electrons which are negative **(1)**; a neutral atom will contain the same number of negative electrons as positive protons **(1)**.

2 A **(1)**

3 (a) When an atom absorbs electromagnetic radiation an electron **(1)** moves to a higher energy level **(1)**.

(b) When an atom emits electromagnetic radiation an electron **(1)** moves to a lower energy level **(1)**.

4 (a) atoms: Li, Cu **(1)**; ions: F⁻, Na⁺, B⁺, K⁺ **(1)**

(b) Atoms are neutral and have no overall charge **(1)**. Ions have gained (−) or lost (+) an electron / have become negatively or positively charged **(1)**.

189. Ionising radiation

1 (a) An alpha particle is a helium nucleus / it is composed of two protons and two neutrons **(1)**, it has a charge of +2 **(1)**.
  (b) A beta particle is an electron **(1)** and; it has a charge of –1 **(1)**.
  (c) A gamma wave is a form of high energy electromagnetic radiation **(1)**; it has no mass or charge **(1)**.

2 B **(1)**

3 alpha: very low, stopped by 10 cm of air **(1)**; beta minus: low, stopped by thin aluminium **(1)**; gamma: very high, stopped by very thick lead **(1)**

4 (a) beta-plus (positron) **(1)**
  (b) alpha particle **(1)**
  (c) neutron **(1)**

5 The process is random. **(1)**

190. Background radiation

1 Radon is a radioactive element **(1)** that is produced when uranium in rocks decays **(1)**.

2 Levels can vary because of the different rocks that occur naturally in the ground **(1)**. They can also vary due to the use of different rocks such as granite in buildings **(1)**.

3 natural: any two from: air, cosmic rays, rocks in the ground, food **(1)**; man-made: any two from: nuclear power, medical treatment, nuclear weapons **(1)**

4 (a) south-east 0.27 Bq **(1)**; south-west 0.30 Bq **(1)**
  (b) south-west **(1)**

5 As the uranium in rocks decays radon gas seeps out **(1)** from the soil and into homes and buildings **(1)**.

191. Measuring radioactivity

1 Badges containing photographic film monitor levels of workers' exposure to radiation, as they become darker over time **(1)**.

2 C, A, D, B (all four correct – **3 marks**, two correct – **2 marks**, one correct – **1 mark**)

3 The student is correct. When radiation is more ionising, it is more likely to create ions **(1)**, so more highly ionising radiation is more likely to ionise the argon in the tube **(1)**, which means that it is more likely to cause a current and be recorded on the rate meter **(1)**.

192. Models of the atom

1 (a) A **(1)**
  (b) The Bohr model showed that electrons **(1)** orbit the atom at different energy levels **(1)** and those electrons have to acquire precise amounts of energy to move up to higher levels **(1)**. The model was an improvement because it was able to explain observed emission and absorption spectra which occur as the atom to be stable **(1)**.

2 Rutherford fired positively charged alpha particles at atoms of gold foil; most went through showing that there were large spaces in the atom **(1)**. Some were repelled or deflected **(1)** showing that the nucleus was positively charged **(1)**.

3 The plum pudding model showed the atom as a 'solid', positively charged **(1)** particle containing a distribution of negatively charged electrons **(1)** whereas the Rutherford model showed the atom as having a tiny, dense, positively charged nucleus **(1)** surrounded by orbiting negatively charged electrons **(1)**.

193. Beta decay

1 β+ particle – a fast-moving positron **(1)**; β– particle – a fast-moving electron **(1)**

2 Beta-minus decay is when a neutron (n) changes to a proton (p) releasing a high-energy electron (e^-) **(1)**. Beta-plus decay is when a proton (p) changes to a neutron (n) releasing a high-energy positron (e^+) **(1)**.

3 (a) 7 **(1)**
  (b) 12 **(1)**

4 (a) In beta-minus decay, a neutron decays into a proton **(1)** and a high-energy electron (beta-minus particle) is emitted **(1)**.
  (b) In beta-plus decay, a proton decays into a neutron **(1)** and a positron (beta-plus particle) is emitted **(1)**.

194. Radioactive decay

1 B and C **(1)**

2 beta-negative, charge (−1) **(1)**; beta-positive, charge (+1) **(1)**

3 The student is wrong because the alpha particle has a mass of 4 nucleons (or 2 protons and 2 neutrons) **(1)** so there is a change in the mass of the nucleus / the nucleus has lost the mass of 4 nucleons (or 2 protons and 2 neutrons) **(1)**.

4 In neutron decay, a neutron **(1)** is emitted and a new isotope of the element is formed **(1)**.

5 B **(1)**

6 (a) (i) Add 208 to Po **(1)**; alpha **(1)**
     (ii) Add 86 to Rn **(1)**; alpha **(1)**
     (iii) Add 42 to Ca **(1)**; beta-minus **(1)**
     (iv) Add 9 to Be **(1)**; neutron **(1)**
  (b) Nucleons often rearrange themselves **(1)** following alpha or beta decay. This causes energy to be emitted as a gamma / photon / wave **(1)**.

195. Half-life

1 The half-life of a radioactive substance is the time it takes **(1)** for half of the (unstable) atoms in a radioactive substance / sample to decay **(1)**.

2 (a) 8 million atoms **(1)**
  (b) 9.3 ÷ 3.1 = 3 half-lives **(1)**, 1 half-life – 8 million; 2 half-lives – 4 million; 3 half-lives – 2 million atoms **(1)**

3 The activity is 500 Bq at 0 minutes **(1)**. Half this activity is 250 Bq, which is at 5 minutes **(1)** so the half-life is 5 minutes. **(1)** (Answers between 4.9 and 5.1 minutes are allowed. If you used other points on the graph and found an answer of around 5 minutes you would get full marks. For this question, your workings can just be pairs of lines drawn on the graph.)

196. Dangers of radiation

1 any two from: hospital **(1)**; dental surgery **(1)**; radiography / X-ray department **(1)**; nuclear power plant **(1)**

2 A **(1)**

3 (a) Ionising means to convert an atom or molecule into an ion, usually by removing one or more electrons. **(1)** (Accept 'electrons are removed from atoms'.)
  (b) Ions in the body can cause damage to cell tissue, **(1)** which can lead to DNA mutations / cancer **(1)**.

4 (a) Employers can limit the time of exposure **(1)**; workers can wear protective clothing / wear a lead apron **(1)**; increase distance from the source **(1)**.
  (b) The amount of energy / dose of radiation that a person has been exposed to is monitored by wearing a film badge **(1)**. This is checked each day **(1)**.

197. Contamination and irradiation

1 Before 1920 the harmful effects of radioactivity were not known / recognised **(1)** so it was thought that it was safe to use **(1)**. It was banned from use once the dangers were known **(1)**.

2 external contamination: radioactive particles come into contact with skin, hair or clothing; internal contamination: a radioactive source is eaten, drunk or inhaled; irradiation: a person becomes exposed to an external source of ionising radiation (all correct for **2 marks**, 1 correct for **1 mark**)

3 (a) any suitable example, e.g. contaminated soil on hands **(1)**
  (b) any suitable example, e.g. contaminated dust or radon gas inhaled **(1)**

4 Internal contamination means that the alpha particles come into contact with the body through inhalation or ingestion **(1)** where they are likely to cause internal tissue damage **(1)**. Irradiation by alpha particles is less likely to cause damage because they have to travel through air **(1)** and are therefore less likely to ionise body cells **(1)** (at distances of over 10 cm).

198. Extended response – Radioactivity

Answer could include the following points: **(6)**
Penetrating ability:
- Alpha is the least penetrating; gamma is the most penetrating.
- Alpha passes through air but not through paper / aluminium / lead.
- Beta passes through air and paper but not aluminium / lead.
- Gamma passes through air / paper / aluminium but not lead.
- Paper stops alpha but not beta or gamma radiation.
- Thin aluminium stops alpha and beta (depending on thickness) but not gamma radiation.
- Very thick lead stops alpha, beta and gamma radiation.

Ionising ability:
- Ionisation happens when an electron is removed from an atom / molecule.
- By collision / interaction with radiation.
- Alpha is the most ionising; gamma is the least ionising.

Answers

- Alpha particles have highest mass / higher mass than beta particles.
- Particles / radiation lose energy when they ionise substances.
- Alpha loses energy fastest / loses most energy so is least penetrating.
- Gamma loses energy most slowly / loses least energy so is most penetrating.

199. Work, energy and power

1 D **(1)**

2 (a) gravitational potential energy store **(1)**
   (b) thermal energy store **(1)**
   (c) chemical energy store **(1)**

3 energy transferred = 15 000 J, time taken = 20 s; $P = E \div t = 15\,000\ \text{J} \div 20\ \text{s}$ **(1)** = 750 **(1)** W **(1)**

4 work done = $F \times d = 600\ \text{N} \times (20 \times 0.08\ \text{m})$ **(1)** = 960 **(1)** J **(1)**

5 $P = E \div t$ so $t = E \div P$ **(1)** = 360 000 J ÷ 200 W **(1)** = 1800 s **(1)** (or 30 minutes)

200. Extended response – Energy and forces

Answer could include the following points: **(6)**

- Kinetic energy is transferred from the moving air particles to the kinetic energy store of the blades of the turbine, as they rotate.
- Kinetic energy is transferred to the sound energy store of the turbine.
- The kinetic energy store of the turbine blades is useful energy.
- The sound and thermal energy store of the turbine is wasted energy.
- The remaining kinetic energy store of the wind is wasted energy.
- Efficiency = useful energy output ÷ total energy input.
- Sankey diagram is sketched to show: input as kinetic energy store of wind, 100 J; outputs as 65 J kinetic energy store of wind and 35 J as combined kinetic energy store of the turbine and thermal energy store of the environment. The arrow for thermal energy should be much smaller than the one for the kinetic energy store of the turbine.
- The mechanical energy of the moving air gives the air particles the ability to apply a force and cause a displacement of the blades.
- Mechanical processes become wasteful when they cause a rise in temperature so dissipating energy to the thermal store of the environment.
- Rise in temperature is caused by friction between moving objects / materials.
- It is important to keep friction as low as possible to minimise wasted energy.
- By reducing wasted energy the wind turbines can be made more efficient.
- Higher efficiency will mean that more electricity is generated.

201. Interacting forces

1 (a) gravitational **(1)**, magnetic **(1)**, electrostatic **(1)**
   (b) A gravitational field is different because it only attracts **(1)** whereas **both** magnetic and electrostatic fields attract and repel **(1)**.

2 A **(1)**, C **(1)**

3 Weight is a vector because it has a direction (towards the centre of the larger mass) **(1)**. Normal contact force is a vector because it has a direction (opposite to the force acting on the surface) **(1)**.

4 (a) pull forward/to the right (by the student on the bag) and friction (between the bag and the floor) opposing direction of motion **(1)** The pull forward / to the right on the bag is larger so the bag moves forward **(1)**.
   (b) weight and normal contact / reaction force **(1)**

202. Circuit symbols

1 so that they can be understood by everybody around the world **(1)**

2 (a) C **(1)**
   (b) (i) The thermistor responds by changing resistance with changes in temperature **(1)**.
       (ii) The LDR responds by changing resistance with changes in light intensity **(1)**.

3

Component	Symbol	Purpose
ammeter	Ⓐ	measures electric current **(1)**
fixed resistor	⊏▭⊐	provides a fixed resistance to the flow of current **(1)**
diode	▷⊳	allows the current to flow one way only **(1)**
switch	⊸⊶ or ⊸⊶	allows the current to be switched on or off **(1)**

203. Series and parallel circuits

1 (a) series: $A_2 = 3\ \text{A}$; $A_3 = 3\ \text{A}$ **(1)**; parallel: $A_2 = 1\ \text{A}$; $A_3 = 1\ \text{A}$; $A_4 = 1\ \text{A}$ **(1)**
   (b) In a series circuit the current is the same throughout the circuit **(1)**. In a parallel circuit the current splits up in each branch **(1)**.

2 (a) series: $V_2 = 3\ \text{V}$ **(1)**; $V_3 = 3\ \text{V}$ **(1)**; $V_4 = 3\ \text{V}$ **(1)**; parallel: $V_2 = 9\ \text{V}$ **(1)**; $V_3 = 9\ \text{V}$ **(1)**; $V_4 = 9\ \text{V}$ **(1)**
   (b) In a series circuit the potential difference is shared / splits up across the components in the circuit **(1)**. In a parallel circuit the potential difference across each branch is the same as the supply potential difference **(1)**.

3 A **(1)**

204. Current and charge

1 (a) An electric current is the rate **(1)** of flow of charge (electrons in a metal) **(1)**.
   (b) $Q = I \times t = 4\ \text{A} \times 8\ \text{s}$ **(1)** = 32 **(1)** coulombs/C **(1)**

2 (a) (i) $A_1 = 0.3\ \text{A}$ **(1)**
       (ii) $A_3 = 0.3\ \text{A}$ **(1)**
   (b) Add another cell / increase the energy supplied **(1)**.
   (c) The electrons move around the circuit in one continuous path **(1)** so the current leaving the cell is the same as the current returning to it **(1)**.

3 (a) any series circuit diagram with a component (e.g. lamp) **(1)** and an ammeter **(1)**
   (b) stopwatch / timer **(1)**

205. Energy and charge

1 Current is the charge flowing per unit time **(1)**. Potential difference is the energy transferred per unit of charge **(1)**.

2 $E = Q \times V$ **(1)** = 30 C × 9 V **(1)** = 270 J **(1)**

3 $E = Q \times V$ so $Q = E \div V$ **(1)** = 125 J ÷ 5 V **(1)** = 25 C **(1)**

4 $Q = I \times t = 0.15\ \text{A} \times 200\ \text{s}$ **(1)** = 30 C; $E = Q \times V = 30\ \text{C} \times 20\ \text{V}$ **(1)** = 600 J **(1)**
   Alternative solution: $E = VIt$ **(1)** so $E = 20\ \text{V} \times 0.15\ \text{A} \times 200\ \text{s}$ **(1)** so $E = 600\ \text{J}$ **(1)**

206. Ohm's law

1 D **(1)**

2 Ohm's law means that the rate of flow of electrons (the current) flowing through the resistor **(1)** is directly proportional to the potential difference across the resistor **(1)**.

3 (a) $R = V \div I = 12\ \text{V} \div 0.20\ \text{A}$ **(1)** = 60 Ω **(1)**
   (b) $R = 22\ \text{V} \div 0.40\ \text{A}$ **(1)** = 55 Ω **(1)**
   (c) $R = 9\ \text{V} \div 0.03\ \text{A}$ **(1)** = 300 Ω **(1)**
   (d) resistor in (c) **(1)**

4 (a) Line A: straight line through origin **(1)**; line B: straight line through origin, different gradient **(1)**.
   (b) the line with the lower gradient **(1)**

207. Resistors

1 A **(1)**

2 (a) potential difference = $I \times R = 2\ \text{A} \times 10\ \Omega$ **(1)** = 20 V **(1)**
   (b) total resistance = sum of the resistances = 10 Ω + 10 Ω **(1)** = 20 Ω **(1)**

3 (a) 20 + 30 + 150 **(1)** = 200 Ω **(1)**
   (b) (i) The sum of the potential differences across the resistors connected in series must equal the potential difference across the battery **(1)**.
       (ii) $I_T = 0.03\ \text{A}$; $R_T = 200\ \Omega$, $V = I \times R = 0.03\ \text{A} \times 200\ \Omega$ **(1)** = 6 V **(1)**; two identical cells so each cell supplies 3 V **(1)**

208. I–V graphs

1 (a) C **(1)**
   (b) As the potential difference increases the current increases **(1)** in a linear / proportional relationship **(1)**.
   (c) As the potential difference increases the current increases **(1)** but the gradient of the line gets less steep / shallower or the increase in current becomes smaller as potential difference continues to increase **(1)**.

2 (a) fixed resistor: same as graph A in Q1 **(1)**; filament lamp: same as graph B in Q1 **(1)**
   (b) The graphs are a different shape from each other because the fixed resistor (at constant temperature) is ohmic / obeys Ohm's law so the current and potential difference have a proportional relationship **(1)**. The filament lamp does

not obey Ohm's law, as temperature increases, so the relationship between current and potential difference is not proportional **(1)**.

3   Use an ammeter to measure current **(1)** and a voltmeter to measure potential difference **(1)**. A variable resistor **(1)** should be included to allow different values of current to be obtained **(1)**. Calculate resistance from Ohm's law **(1)**.

209.   Electrical circuits

1   (a) two resistors in same loop **(1)**; at least one ammeter shown and connected in series **(1)**; at least one voltmeter shown and connected in parallel across a resistor or cell/battery **(1)**

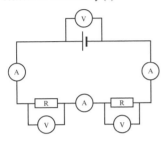

(b) two resistors in separate loops **(1)**; at least one ammeter shown and connected in main circuit or in a loop connected in series **(1)**; at least one voltmeter shown and connected in parallel across a resistor or cell / battery **(1)**

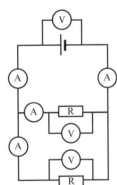

(c) Current is the same at any point in a series circuit **(1)** but will split up at a junction in a parallel circuit **(1)**. The sum of the potential difference across components in a series circuit equals the potential difference of the cell **(1)**. The sum of the potential difference across components in each loop in a parallel circuit equals the potential difference of the cell **(1)**.

2   (a) Connect a cell, a filament lamp, a variable resistor and an ammeter in a series circuit **(1)**. Connect the voltmeter across the filament lamp in parallel **(1)**. Adjust the variable resistor setting to obtain a number of readings for current and potential difference **(1)**.

(b) Ohm's law / resistance = potential difference ÷ current ($R = V \div I$) **(1)**

(c) Plotting a graph of I against V shows how the current through it varies with the potential difference across it. The gradient of the I–V graph is equal to $1 / R$ **(1)**, so inverting the value of the gradient gives you the resistance, R **(1)**.

210.   The LDR and the thermistor

1

Light-dependent resistor (LDR)	Thermistor
⌽ **(1)**	⌿ **(1)**

2   D **(1)**

3   (a) The resistance goes down (more current flows) as the light becomes more intense (brighter) **(1)**.

(b) The resistance goes down (more current flows) as the temperature goes up **(1)**.

4   The lamp lights up when the temperature is high **(1)** because the current through the lamp and the thermistor will be high when the resistance of the thermistor falls **(1)**.

5   When the level of light increases, the resistance decreases **(1)** and the current increases **(1)**.

211.   Current heating effect

1   A **(1)**

2   any three suitable examples, e.g. electric fire **(1)**, hairdryer **(1)**, kettle **(1)**, iron **(1)**, toaster **(1)**

3   When a conductor is connected to a potential difference the free electrons **(1)** move through the lattice of metal ions **(1)**. As they do so, collisions **(1)** occur where the kinetic energy is transferred into thermal energy **(1)**, causing the heating effect.

4   The appliances listed each need a lot of current. **(1)** When these are all added to a single socket they will, together, draw a high current. **(1)** High current results in a heating effect in the wires, **(1)** which could lead to a fire **(1)**.

212.   Energy and power

1   B **(1)**

2   (a) using the equation for power $P = I \times V$ = 5 A × 230 V **(1)** = 1150 **(1)** W

(b) $E = I \times V \times t$ = 0.2 A × 4 V × 30 s **(1)** = 24 **(1)** J **(1)**

3   (a) $P = I \times V$ so $I = P \div V$ **(1)** = 3 W ÷ 6 V **(1)** = 0.5 A **(1)**

(b) $E = I \times V \times t$ = 0.5 A × 6 V × 300 s **(1)** = 900 **(1)** J **(1)** OR $E = P \times t$ = 3 W × 300 s **(1)** = 900 **(1)** J **(1)**

213.   a.c. and d.c. circuits

1   (a) Alternating current is an electric current that changes direction regularly **(1)** and its potential difference is constantly changing **(1)**.

(b) Direct current is an electric current in which all the electrons flow in the same direction **(1)** and its potential difference has a constant value **(1)**.

2   D **(1)**

3   (a) $P = E \div t$, so $E = P \times t$, so E = 2000 W × (15 × 60) s **(1)** = 1 800 000 **(1)** J

(b) E = 1500 W × 25 s **(1)** = 37 500 **(1)** J

(c) E = 10 W × (6 × 60 × 60) s **(1)** = 216 000 **(1)** J

214.   Mains electricity and the plug

1   (a) one correct **one mark**; three correct **two marks**: earth wire (green and yellow); live wire (brown); neutral wire (blue); fuse

(b) The fuse is connected to the live wire **(1)** because the live wire carries the current into the appliance **(1)**.

2   brown: electrical current enters the appliance at 230 V **(1)**; blue: electrical current leaves the appliance at 0 V through this wire **(1)**; green/yellow: this is a safety feature connected to the metal casing of the appliance **(1)**.

3   When a large current enters the live wire **(1)** this transfers to the thermal energy store of the wire **(1)** which melts the wire in the fuse **(1)**. The circuit is then broken **(1)**.

4   (a) When a current is too high **(1)** a strong magnetic field is generated which opens a switch (held back by a spring) **(1)**. This 'breaks' the circuit **(1)**, making it safe.

(b) The earth wire is connected to the metal casing **(1)**. If the live wire becomes loose and touches anything metallic **(1)** the current passes through the earth wire instead of electrifying the casing or metal components / causing a shock **(1)**.

215.   Extended response – Electricity and circuits

Answer could include the following points: **(6)**

- The thermistor can be connected in series with an ammeter to measure current with a voltmeter connected in parallel across it to measure potential difference.
- Ohm's law can be referred to in calculating the resistance.
- When the temperature is low the resistance of the thermistor will be high, allowing only a small current to flow.
- When the temperature is high the resistance of the thermistor will be low, allowing a larger current to flow.
- The light-dependent resistor can be connected in series with an ammeter to measure current with a voltmeter connected in parallel across it to measure potential difference.
- When light levels are low (dark) the resistance of the light-dependent resistor will be high, allowing only a small current to flow.
- When light levels are high (bright) the resistance of the light-dependent resistor will be low, allowing a larger current to flow.
- Thermistors can be used in fire alarms as a temperature sensor to switch on an alarm.
- Light-dependent resistors can be used in security systems as a light sensor to switch on a light.

216.   Magnets and magnetic fields

1   (a) bar magnet: field line out (arrows) at N **(1)**, field lines in (arrows) at S **(1)**, field line close at poles **(1)**, further apart at sides **(1)**

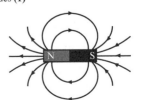

(b) uniform field: parallel field lines **(1)**, arrows from N to S **(1)**

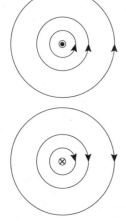

2   Both a bar magnet and the Earth have north and south poles **(1)**. They also both have similar magnetic field patterns **(1)**.

3   A temporary magnet is used for an electric doorbell because it can be magnetised when the current is switched on **(1)**, which attracts the soft iron armature to ring the bell **(1)**, and de-magnetised when the current is switched off **(1)** (returning the armature away from the bell).

4   Rajesh can do a second test by moving a permanent magnet near the magnetic materials **(1)**. Those that are attracted but not repelled will be temporary magnets **(1)**. The materials that can be attracted and repelled are permanent magnets **(1)**.

217.   Current and magnetism

1   (a) at least two concentric circles on each diagram **(2)**
    (b) clockwise arrows on cross diagram **(1)**, anticlockwise arrows on dot diagram **(1)**

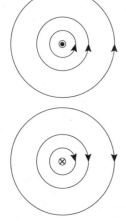

2   B **(1)**

3   (a) The strength of the magnetic field depends on the size of the current in the wire **(1)** and the distance from the wire **(1)**.
    (b) left: x-axis marked 'Current' **(1)**
        right: x-axis marked 'Distance from the wire' **(1)**

4   (a) (i) increase **(1)**
        (ii) decrease **(1)**
    (b) (i) decrease **(1)**
        (ii) increase **(1)**

218.   Extended response – Magnetism and the motor effect

Answer could include the following points: **(6)**

* A long straight conductor could be connected to a cell, an ammeter and a small resistor to prevent overheating in the conductor.
* When the current is switched on the direction of the magnetic field generated

around a long straight conductor can be found using the right-hand grip rule.

* The right-hand grip rule points the thumb in the direction of conventional current and the direction of the fingers shows the direction of the magnetic field.
* A card can be cut halfway through and placed at right angles to the long straight conductor. A plotting compass can be used to show the shape and direction of the magnetic field.
* The shape of the magnetic field around the long straight conductor will be circular / concentric circles as the current flows through it.
* The strength of the magnetic field depends on the distance from the conductor.
* The concentric magnetic field lines mean that the field becomes weaker with increasing distance.
* The strength of the magnetic field can be increased by increasing the current.

219.   Transformers

1   A step-up transformer is used to increase voltage and decrease current **(1)**. A step-down transformer is used to decrease voltage and increase current **(1)**.

2   (a) As the voltage is increased, the current goes down **(1)** so this reduces the heating effect due to resistance **(1)** and means that less energy is wasted in transmission **(1)**.
    (b) The voltages are high enough to kill you if you touch or come into contact with a transmission line **(1)**.

3   step-down transformer: decreases voltage **(1)**

4   $V_P \times I_P = V_S \times I_S$ so rearranging gives $V_S = V_P \times I_P \div I_S$ **(1)** = 300 V × 0.5 A ÷ 10 A **(1)** = 15 **(1)** V **(1)**

220.   Extended response – Electromagnetic induction

Answer could include the following points: **(6)**

* Electrical energy is transferred at high voltages from power stations to reduce the current.
* A lower current results in reduced heating effect, and so reduced energy loss from the cables, during transmission.
* Step-up transformers are necessary to increase the voltage of the electrical energy transmitted to the National Grid.
* Voltage is increased due to electromagnetic induction; according to the transformer power equation voltage in the primary coil × current in the primary coil = voltage in the secondary coil × current in the secondary coil ($V_P \times I_P = V_S \times I_S$).
* Transformers are assumed to be 100% efficient.
* Step-down transformers are necessary to decrease the voltage of the electrical energy transmitted to domestic supplies to a safer level.

221.   Changes of state

1   Particles move around each other – liquid – some intermolecular forces. Particles

cannot move freely – solid – strong intermolecular forces. Particles move randomly – gas – almost no intermolecular forces. (All four links correct for **2 marks**, one correct for **1 mark**.)

2   (a) They all consist of particles **(1)**.
    (b) Particles have different amounts in the kinetic energy store **(1)** and experience different intermolecular forces **(1)**.

3   B **(1)**

4   At boiling point the liquid changes state **(1)** so the energy applied after boiling point is reached goes into breaking bonds **(1)** between the liquid particles. The particles gain more energy and become a gas **(1)**.

5   The kinetic energy **(1)** of the particles decreases **(1)** as the ice continues to lose energy to the surroundings. This is measured as a fall in temperature **(1)**.

222.   Density

1   (a) $\rho = m \div V$ = 1650 kg ÷ 3 m³ **(1)** = 550 kg/m³ **(1)**;
    (b) $\rho = m \div V$ **(1)** = 4 kg ÷ 0.005 m³ **(1)** = 800 kg/m³ **(1)**

2   A **(1)**, B **(1)**

3   volume = 0.1 m × 0.25 m × 0.15 m = 0.003 75 m³. $\rho = m \div V$ so $m = \rho \times V$ **(1)** = 3000 kg/m³ × 0.003 75 m³ **(1)** = 11.25 kg **(1)**

4   Marco has approached this problem by stating a scientific principle **(1)** relating density to states of matter but he has not tried to investigate this **(1)**. Ella has approached this problem by observing and comparing **(1)** different densities, but she has not tried to explain this **(1)**. Both students should expand their approach so that observations are explained through scientific principles **(1)**.

223.   Investigating density

1   (a) mass **(1)**
    (b) electronic balance **(1)**

2   (a) The volume of mass may be found by measuring its dimensions **(1)** or by using a displacement method, such as immersing the object in water in a measuring cylinder, to measure how much liquid the mass displaces **(1)**.
    (b) The measurement method is suitable for regular-shaped objects **(1)** whereas the displacement method is best for irregular-shaped objects, where measuring dimensions would be more difficult **(1)**.

3   (a) Place a measuring cylinder on a balance and then zero the scales with no liquid present in the measuring cylinder **(1)**. Add the liquid to the required level **(1)**. Record the mass of the liquid (in g) from the balance and its volume (in cm³) from the measuring cylinder **(1)**.
    (b) Take the value at the bottom of the meniscus **(1)**. Make sure that the reading is taken with the line of sight from the eye to the meniscus perpendicular to the scale to avoid a parallax error **(1)**.
    (c) density = mass ÷ volume = 121 g ÷ 205 cm³ = 0.59 **(1)** g/cm³ **(1)**

224. Energy and changes of state

1 Specific latent heat is the energy transfer needed to change the state of 1 kg of a substance with no rise in temperature **(1)**.

2 $\Delta Q = m \times c \times \Delta\theta$ **(1)** = 0.8 kg × 4200 J/kg °C × 50 °C **(1)** = 168 000 J **(1)**

3 $Q = m \times L$ = 35 kg × 336 000 J/kg **(1)** = 11 760 000 J **(1)**

4 (a) 'gas' in top right box; 'solid' in bottom box **(1)** (both needed for mark); 'evaporating' in box by top horizontal line; 'melting' in box by bottom horizontal line **(1)**
   (b) The energy being transferred to the material is breaking bonds **(1)** and as a result, the material undergoes a phase change **(1)**.

5 total energy needed = energy to heat water from 20 °C to 100 °C + energy to turn water at 100 °C into steam; energy to heat water = $m \times c \times \Delta\theta$ = 0.5 kg × 4200 J/kg °C × (100 − 20) °C **(1)** = 168 000 J **(1)**; energy to turn water into steam = $m \times L$ = 0.5 kg × 2 265 000 J/kg **(1)** = 1 132 500 J **(1)**; total energy = 168 000 J + 1 132 500 J = 1 300 500 J **(1)**

225. Thermal properties of water

1 (a) the amount of energy required to raise the temperature of 1 kg of material by 1 K (or 1 °C) **(1)**
   (b) specific heat capacity = change in thermal energy ÷ (mass × change in temperature) or ($c = \Delta Q \div (m \times \Delta\theta)$ **(1)**

2 (a) Place a beaker on a balance, zero the balance and add a measured mass of water **(1)**. Take a start reading of the temperature **(1)**. Place the electrical heater into the water and switch on **(1)**. Take a temperature reading every 30 seconds (or suitable time interval) **(1)** until the water reaches the required temperature **(1)**.
   (b) Measure the current supplied, the potential difference across the heater and the time for which the current is switched on **(1)**. Use these values to calculate the thermal energy supplied using the equation $E = V \times I \times t$ **(1)**.
   (c) Add insulation around the beaker **(1)** so less thermal energy is transferred to the surroundings and a more accurate value for the specific heat capacity of the water may be obtained **(1)**.

3 Plot a graph of temperature against time **(1)**. The changes of state are shown when the graph is horizontal (the temperature is not increasing) **(1)**.

4 Both experiments use an electrical heater close to water so there is a danger of electric shock – keep electrical wires and switches dry **(1)**. Both experiments use water that could be spilled and cause slippage – report and wipe up immediately **(1)**. (Note: specific latent heat experiments tend not to use glass beakers (which could break and cause cuts in the specific heat capacity experiment) but tend to use metal containers, so glass is not necessarily common to both experiments. The specific heat capacity experiment does not require water to be heated to a level to cause scalds so the hot water / water vapour hazard in the specific latent heat experiment is not necessarily common to both experiments.)

226. Pressure and temperature

1 Temperature is a measurement of the average kinetic energy of the particles in a material **(1)**.

2 (a) 273 K → 0 °C **(1)**, 255 K → −18 °C **(1)**, 373 K → 100 °C **(1)**
   (b) (i) At 0K, the kinetic energy **(1)** of the particles would be zero **(1)**. 0K is known as 'absolute zero' **(1)**.
       (ii) −273 °C **(1)**

3 (a) As the temperature increases the particles will move faster **(1)** because they gain more energy **(1)**.
   (b) As the particles are moving faster they will collide with the container walls more often **(1)** therefore increasing the pressure **(1)**.
   (c) It increases. **(1)**

4 The kinetic energy of the particles will also increase by a factor of four **(1)** because temperature and average kinetic energy are directly proportional **(1)**.

227. Extended response – Particle model

Answer could include the following points: **(6)**

The transfer of energy from the thermal energy stores may be reduced in each of the following areas:

- Roof: loft insulation traps air (insulator) reducing energy transfer from the thermal energy store of the roof by conduction (vibration of solid particles) through room ceiling.
- Walls: cavity wall insulation (insulating) foam traps air (insulator) reducing energy transfer from the thermal energy store of the walls by conduction (vibration of solid particles).
- Floor: carpets (insulating material) trap air (insulator) reducing energy transfer from the thermal energy store of the air to the floor by conduction (vibration of solid particles).
- Doors: fit draught excluders to reduce the cooling effect from draughts reducing energy transfer from the thermal energy store of the air by convection (movement of fluid particles) around room.
- Windows: the space between the panes of glass in double (or triple) glazing reduces energy transfer from the thermal energy store of the glass by conduction (vibration of solid particles).
- Windows: heavy/thick curtains (insulating material) reduce energy transfer from the thermal energy store of the air to the glass by conduction (vibration of solid particles).

228. Elastic and inelastic distortion

1 push forces (towards each other): compression **(1)**, pull forces (away from each other): stretching **(1)**, clockwise and anticlockwise: bending **(1)**

2 (a) washing line (or any valid example) **(1)**
   (b) G-clamp, pliers (or any valid example) **(1)**
   (c) fishing rod (with a fish on the line) (or any valid example) **(1)**
   (d) dented can or deformed spring (or any valid example) **(1)**

3 After testing, beam 1 would return to the same size and shape as prior to the test **(1)** under the load and would be intact **(1)**. Beam 2 would distort and change shape **(1)** but would (probably) still be intact **(1)**.

4 Crumple zones use inelastic distortion **(1)** and are designed to distort / change shape **(1)** in the event of a crash. They extend the time taken for a body to come to rest, reducing the force on the body **(1)**.

229. Springs

1 Elastic means that the object will return to original size and shape **(1)** (*both needed for mark*) after the deforming force is removed **(1)**.

2 extension = 0.07 m − 0.03 m = 0.04 m; force = spring constant / k × extension = 80 N × 0.04 m **(1)** = 3.2 **(1)** N **(1)**

3 D **(1)**

4 (a) $F = k \times x$ = 200 N/m × 0.15 m **(1)** = 30 N **(1)**
   (b) $E = \frac{1}{2} \times k \times x^2 = \frac{1}{2} \times 200$ N/m × (0.15 m)^2 **(1)** = 2.25 J **(1)**

230. Forces and springs

1 (a) Hang a spring from a clamp attached to a retort stand and measure the length before any masses or weights are added using a half-metre ruler, marked in mm **(1)**. Carefully add the first mass or weight and measure the total length of the extended spring **(1)**. Unload the mass or weight and re-measure the spring to make sure that the original length has not changed **(1)**. Add at least five masses or weights and repeat the measurements each time **(1)**.
   (b) The spring only stores elastic potential energy when it can all be recovered **(1)** and is not transferred to cause a permanent change of shape in the spring **(1)**.
   (c) The extension of the spring must be calculated for each force by taking away the original length of the spring from each reading **(1)**. Extension measurements should be converted to metres **(1)**.
   (d) (i) The area under the graph equals the work done / the energy stored in the spring as elastic potential energy **(1)**.
       (ii) The gradient of the linear part of the force–extension graph gives the spring constant k **(1)**.
   (e) energy stored = $\frac{1}{2} \times k \times x^2$ **(1)**

2 The length of a spring is measured with no force applied to the spring whereas the extension of a spring is the length of the spring measured under load / force less the original length **(1)**.

231. Extended response – Forces and matter

Answer could include the following points: **(6)**

- Two or more forces are required to cause an object to distort / deform / change shape as one force has to hold the object in position.
- Elastic distortion results in the object returning to its original shape whereas inelastic distortion results in a permanent change of shape.
- Energy transferred to the object by a force can be stored if it deforms elastically and the force maintained. When the force is removed, energy may be recovered from the potential elastic / spring energy store of the object.

- When an object permanently changes shape, the energy transferred to the object by a force cannot be recovered as it has, instead, caused permanent distortion / changed of shape.
- Metal springs usually exhibit elastic distortion when the force is below that which would cause a permanent change of shape.
- The relationship between force and the extension of a spring is usually directly proportional (if the force doubles, the extension doubles) during elastic distortion / change of shape.
- Elastic distortion may be useful in sports events, such as diving, archery or pole

vaulting, where the potential energy stored can be recovered and used by the athlete (any valid example).

- Plastic distortion may be useful when producing industrial goods, such as car panels or plastic bottles, where a force results in a permanent change of shape (any valid example).

The Periodic Table of the Elements

Key:
- relative atomic mass
- **atomic symbol**
- name
- atomic (proton) number

Example:
1
H
hydrogen
1

Groups: 1, 2, 3, 4, 5, 6, 7, 0

1	2											3	4	5	6	7	0
																	4 **He** helium 2
7 **Li** lithium 3	9 **Be** beryllium 4											11 **B** boron 5	12 **C** carbon 6	14 **N** nitrogen 7	16 **O** oxygen 8	19 **F** fluorine 9	20 **Ne** neon 10
23 **Na** sodium 11	24 **Mg** magnesium 12											27 **Al** aluminium 13	28 **Si** silicon 14	31 **P** phosphorus 15	32 **S** sulfur 16	35.5 **Cl** chlorine 17	40 **Ar** argon 18
39 **K** potassium 19	40 **Ca** calcium 20	45 **Sc** scandium 21	48 **Ti** titanium 22	51 **V** vanadium 23	52 **Cr** chromium 24	55 **Mn** manganese 25	56 **Fe** iron 26	59 **Co** cobalt 27	59 **Ni** nickel 28	63.5 **Cu** copper 29	65 **Zn** zinc 30	70 **Ga** gallium 31	73 **Ge** germanium 32	75 **As** arsenic 33	79 **Se** selenium 34	80 **Br** bromine 35	84 **Kr** krypton 36
85 **Rb** rubidium 37	88 **Sr** strontium 38	89 **Y** yttrium 39	91 **Zr** zirconium 40	93 **Nb** niobium 41	96 **Mo** molybdenum 42	[98] **Tc** technetium 43	101 **Ru** ruthenium 44	103 **Rh** rhodium 45	106 **Pd** palladium 46	108 **Ag** silver 47	112 **Cd** cadmium 48	115 **In** indium 49	119 **Sn** tin 50	122 **Sb** antimony 51	128 **Te** tellurium 52	127 **I** iodine 53	131 **Xe** xenon 54
133 **Cs** caesium 55	137 **Ba** barium 56	139 **La*** lanthanum 57	178 **Hf** hafnium 72	181 **Ta** tantalum 73	184 **W** tungsten 74	186 **Re** rhenium 75	190 **Os** osmium 76	192 **Ir** iridium 77	195 **Pt** platinum 78	197 **Au** gold 79	201 **Hg** mercury 80	204 **Tl** thallium 81	207 **Pb** lead 82	209 **Bi** bismuth 83	[209] **Po** polonium 84	[210] **At** astatine 85	[222] **Rn** radon 86
[223] **Fr** francium 87	[226] **Ra** radium 88	[227] **Ac*** actinium 89	[261] **Rf** rutherfordium 104	[262] **Db** dubnium 105	[266] **Sg** seaborgium 106	[264] **Bh** bohrium 107	[277] **Hs** hassium 108	[268] **Mt** meitnerium 109	[271] **Ds** darmstadtium 110	[272] **Rg** roentgenium 111							

Elements with atomic numbers 112–116 have been reported but not fully authenticated

*The lanthanoids (atomic numbers 58–71) and the actinoids (atomic numbers 90–103) have been omitted.

The relative atomic masses of copper and chlorine have not been rounded to the nearest whole number.

259

Physics Equations List

(final velocity)^2 − (initial velocity)^2 = 2 × acceleration × distance

$v^2 - u^2 = 2 \times a \times x$

energy transferred = current × potential difference × time

$E = I \times V \times t$

change in thermal energy = mass × specific heat capacity × change in temperature

$\Delta Q = m \times c \times \Delta\theta$

thermal energy for a change of state = mass × specific latent heat

$Q = m \times L$

energy transferred in stretching = 0.5 × spring constant × (extension)^2

$E = \frac{1}{2} \times k \times x^2$